SpringerBriefs in Electrical and Computer Engineering

SpringerBriefs present concise summaries of cutting-edge research and practical applications across a wide spectrum of fields. Featuring compact volumes of 50 to 125 pages, the series covers a range of content from professional to academic. Typical topics might include: timely report of state-of-the art analytical techniques, a bridge between new research results, as published in journal articles, and a contextual literature review, a snapshot of a hot or emerging topic, an in-depth case study or clinical example and a presentation of core concepts that students must understand in order to make independent contributions.

More information about this series at http://www.springer.com/series/10059

Xiaodong Lin • Jianbing Ni
Xuemin (Sherman) Shen

Privacy-Enhancing Fog Computing and Its Applications

Xiaodong Lin
Department of Physics and Computer Science
Wilfrid Laurier University
Waterloo, ON, Canada

Jianbing Ni
Electrical and Computer Engineering Department
University of Waterloo
Waterloo, ON, Canada

Xuemin (Sherman) Shen
Electrical and Computer Engineering Department
University of Waterloo
Waterloo, ON, Canada

ISSN 2191-8112 ISSN 2191-8120 (electronic)
SpringerBriefs in Electrical and Computer Engineering
ISBN 978-3-030-02112-2 ISBN 978-3-030-02113-9 (eBook)
https://doi.org/10.1007/978-3-030-02113-9

Library of Congress Control Number: 2018959867

This Springer imprint is published by the registered company Springer Nature Switzerland AG
The registered company address is: Gewerbestrasse 11, 6330 Cham, Switzerland

Preface

Fog computing has been considered as a key enabler to reach the increasing demands on local data analysis and numerous device connections for Internet-of-Things (IoT), by extending computing, storage, and networking resources to the network edge. Deployed between user devices and cloud centers, fog nodes monitor or analyze real-time data from network-connected "things," supporting a variety of IoT services, such as smart traffic lights, home energy management, and augmented reality. As a promising extension of cloud computing, fog computing can offer on-demand and ubiquitous applications on nearby devices that can result in superior user experience and increase redundancy in case of failures. Despite the appealing advantages, fog computing is confronted with various security and privacy threats due to ubiquitous connections and limited resources, which have not been systemically discussed in the literature.

In this monograph, we address the security and privacy challenges in fog computing and propose secure and privacy-preserving schemes to deal with these challenges for securing fog-assisted IoT applications. The research is of great importance since security and privacy problems faced by fog computing impede the healthy development of its enabled IoT applications. In Chap. 1, we introduce the architecture of fog-assisted IoT applications and the security and privacy challenges in fog computing. In Chap. 2, we review several promising privacy-enhancing techniques and show examples on how to leverage these techniques to enhance the privacy of users in fog computing. Specifically, we divide the existing privacy-enhancing techniques into three categories: identity privacy-enhancing techniques, location privacy-enhancing techniques, and data privacy-enhancing techniques. With the advanced privacy-enhancing techniques, we propose three secure and privacy-preserving schemes for fog computing applications, including smart parking navigation, mobile crowdsensing, and smart grid, which will be detailed in the next three chapters. In Chap. 3, we introduce the identity privacy leakage in smart parking navigation systems and propose a privacy-preserving smart parking navigation scheme to prevent identity privacy exposure and support efficient parking guidance retrieval through road-side units (fogs) with high retrieving probability and security guarantees. In Chap. 4, we introduce the location privacy leakage during task allocation in mobile crowdsensing and propose a strong privacy-preserving task allocation scheme that enables location-based task allocation based on fog computing without exposing knowledge about the location of participators in mobile crowdsensing. In Chap. 5, we introduce the data privacy leakage in smart grid and propose an efficient and privacy-preserving smart metering scheme to allow collectors (fogs) to achieve real-time measurement collection with privacy-enhanced data aggregation, even if the collectors are compromised or curious about collected data. Finally, remarks and future research directions are given in Chap. 6. This monograph validates the significant feature extension and efficiency improvement of IoT devices without sacrificing the security and privacy of users against dishonest fog nodes. It also provides valuable insights on the security and privacy protection for fog-enabled IoT applications.

We would like to thank Prof. Kuan Zhang at the University of Nebraska-Lincoln and Prof. Yong Yu at Shaanxi Normal University for their contributions in the presented research works. We would also like to thank Dongxiao Liu, Cheng Huang, Meng Li, and Liang Xue for reviewing parts of this monograph, and all the members of broadband communication research group for the valuable discussions and their insightful suggestions and comments. Special thanks are also due to the staff at Springer Science+Business Media, especially Susan Lagerstrom-Fife for her help throughout the publication preparation process.

Waterloo, Canada

Xiaodong Lin
Jianbing Ni
Xuemin (Sherman) Shen

Contents

Acronyms

AM	Auction Manager
AMRC	Advanced Metering Regional Collectors
BBS	Bulletin Board System
BIDH	Computational Bilinear Inverse Diffie-Hellman
CA	Certificate Authority
CONF	Conference-Key Sharing
DDH	Decisional Diffie-Hellman Problem
DDoS	Distributed Denial-of-Service
DHI	Diffie-Hellman Inversion
DL	Discrete Logarithm
DMV	Department of Motor Vehicles
DSRC	Dedicated Short Range Communications
FaaS	Fog as a Service
GPS	Global Positioning System
IDS	Intrusion Detection System
IoT	Internet of Things
IPS	Intrusion Prevention System
LBS	Location-Based Service
LTE	Long-Term Evolution
MCS	Mobile Crowdsensing
MM	Membership Manager
MPC	Multi-Party Computation
OBU	Onboard Unit
OC	Operation Center
P-SPAN	Privacy-preserving Smart Parking Navigation System
PKE	Public-Key Encryption
PKI	Public Key Infrastructure
PoS	Proof-of-Stake
PoW	Proof-of-Work
PPFMC	Privacy-Preserving Fog-assisted Mobile Crowdsensing
P^2SM	Privacy-Preserving Smart Metering

RFID	Radio-Frequency Identification
RSU	Roadside Unit
SNARG	Succinct Non-interactive Arguments
TA	Trusted Authority
TM	Tracing Manager
V2I	Vehicle-to-Infrastructure
V2V	Vehicle-to-Vehicle
VANET	Vehicular Ad Hoc Network
ZKPoK	Zero-Knowledge Proof-of-Knowledge

Chapter 1
Introduction

1.1 Fog Computing

A large number of physical "things", embedded with sensors and actuators, exchange data with each other or the Internet through heterogeneous networks, which brings us to the era of Internet of Things (IoT) [1]. Currently, various devices, such as smart phones, appliances, traffic lights, wearable devices, vehicles and industrial sensors, are interconnected to offer a variety of services and applications in different domains, including smart city, e-healthcare, intelligent transportation and disaster response [2]. This innovation has brought new opportunities to improve our lives and promote the development of our society.

With the connection of numerous devices, massive data is produced from diverse services and applications, resulting in overwhelming pressure on data storage, processing and analysis. It is predicted that the devices connected to the Internet would produce 507.5 ZB per year by 2019. Forwarding these data from the devices to the cloud will require huge communication bandwidth, storage spaces and computing resources [3, 4]. However, 45 percent of produced data will be maintained and processed locally, i.e., at the network edge or near the devices. Furthermore, all the produced data should be analyzed on time to discover the knowledge for satisfying the application demands. For example, the data collected by self-driving vehicles should be processed locally for real-time decision. Thereby, centralized data storage and analysis cannot afford the demands of data-driven applications and latency of service responses [5].

Fog computing [6] has been introduced to enable data-driven services and latency-sensitive applications, by pushing storage and network resources to the network edge. Towards conventional communication networks, the evolution significantly improves the connection capability and network resources utility between the end devices and the cloud. By utilizing the decentralized resources, fogs, which can be macro/small cell base stations and Wi-Fi hotspots with extended computing,

X. Lin et al., *Privacy-Enhancing Fog Computing and Its Applications*,
SpringerBriefs in Electrical and Computer Engineering,
https://doi.org/10.1007/978-3-030-02113-9_1

storage and network capabilities, cooperatively handle a large amount of tasks on communication, storage, computation and management to improve the efficiency of data collection, forwarding and analytics [7]. The fog computing brings various appealing conveniences to IoT services, including low latency, high bandwidth, location awareness and local real-time service [8].

Fog computing is defined by Cisco in 2012 as an extension of cloud computing that offers computation, storage, and networking services between end devices and cloud servers [9]. It does not replace the cloud, but complements it; fogs facilitate a hierarchical infrastructure, in which transit data storage and local data analysis are performed at fogs, and permanent storage and global analysis occur at the cloud. With basic computing, storage and networking resources, users can lease the facilities and resources to access new services and applications. "Fog as a Service (FaaS)" [10] has become an exciting new opportunity, in which a fog service provider builds an array of fogs at geographic locations to offer certain services from vertical markets. Thereby, fog computing is deemed as a service model, where data are stored, processed and analyzed within the network, rather than in a centralized cloud [11].

Fog computing deals with the increasing number of connected devices (things) and emerging applications in IoT by smartly orchestrating and managing computing, storage and networking resources provisioned at the edge of network [12]. By utilizing the resources close to end users, fog computing offers a series of novel applications and services, such as hierarchical data analysis, smart traffic lights, smart wind farms and smart parking reservation, and assist to resolve challenges of high delay and constrained bandwidth of cloud-based services and limited resources of IoT devices. In fog-enabled IoT, the time-sensitive data are maintained and processed on fogs that are close to the devices for real-time control and analytics [13]. Fogs periodically forward data summaries to the cloud, if permanent storage and global analysis are needed. Hence, fog computing is not a competitor of the cloud in IoT; on the contrary, it is envisioned as a perfect complement for large amounts of applications, where cloud computing is insufficient to satisfy some particular demands.

1.2 Fog-Enabled IoT

Fog computing can extend a variety of IoT applications under the architecture of fog-enabled IoT applications.

1.2.1 Architecture of Fog-Enabled IoT

The adaptive and systematic integration of fog computing and IoT expands network connectivity and capability to support nearly omnipotent and ubiquitous IoT services. With the network sources at the edge of networks, fogs localize IoT services and data storage on behalf of the intermediates, connecting the upper cloud layer and the button device layer, as given in Fig. 1.1.

Cloud–The cloud extends various IoT applications for devices with nearly unlimited storage and computing resources. The data centers are responsible for offering outsourced data storage services to devices and computing capabilities for data analytics. The application servers tackle the access requests from devices and utilize rich network, computing and storage resources to offer data-intensive IoT services.

Fog–The fog layer, a network of fogs, is composed of a large amount of macro/small cell base stations and Wi-Fi hotspots. In addition to their ability to build network connections between devices and cloud servers, they are also able to employ computing and storage resources to provide local services and IoT applications. According to the recent wireless technologies, such as 5G, LTE, mmWave, DSRC, and Wi-Fi, fogs achieve up-link and down-link data transmissions with reduced communication overhead. Intuitively, fogs are extended from traditional base stations and hotspots with computing and storage capabilities to support data pre-processing and caching for IoT and low-latency services to devices.

Devices–Massive devices are connected to the cloud for various IoT applications. There are two types of IoT devices, mobile devices and stationary devices. Mobile devices are carried by their owners, such as fitness trackers, smart clothes, smart phones, wearable cameras, smart glasses, smart watches and vehicles, whereas the stationary devices, including environmental sensors and RFID tags, which are pre-deployed at certain areas or on particular products. Smart devices have the

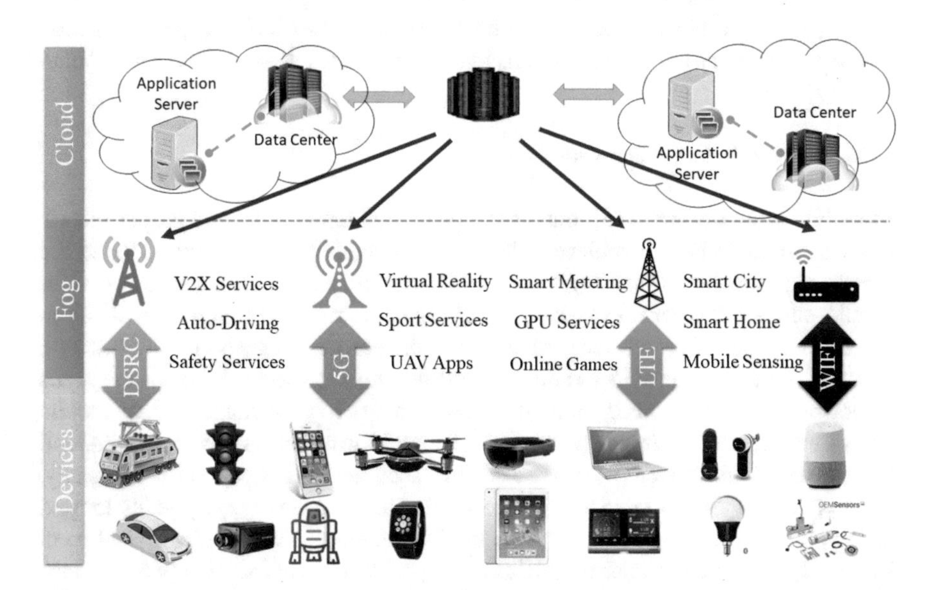

Fig. 1.1 Architecture of fog-enabled IoT applications

capabilities to obtain their interested data from environment, deliver produced data to cloud servers, and access latency-sensitive services.

1.2.2 Applications of Fog-Enabled IoT

We briefly introduce several fog-enabled IoT applications, namely, smart parking navigation, mobile crowdsensing and smart grid, which rely on the local data processing for service optimization.

1.2.2.1 Smart Parking Navigation

Finding an appropriate driving route to a desired destination in a congested area or an unfamiliar region is time-consuming and frustrating for drivers. Real-time traffic information is critical for congestion monitoring and vehicle navigation for drivers. Collecting real-time road conditions and proper paths response for drivers in a timely fashion are challenging problems in vehicular navigation systems [14]. Fog computing can play an essential role in local data collection and navigation result response. To be specific, fogs, which are upgraded roadside units that stretch to have computing capability and storage spaces, can maintain traffic information collected from the driving vehicles in their coverage areas. If a fog receives a navigation request from a vehicle, it can cooperate with other fogs to generate a proper driving path for the querying vehicle to its desirable destination, and rapidly returns the path to the querying vehicle. In this way, the vehicle enjoys flexible real-time navigation services and can promptly take actions to avoid being stuck in traffic congestion.

1.2.2.2 Mobile Crowdsensing

In mobile crowdsensing [15], one of the main challenges is to find proper mobile users for spatial tasks to achieve efficient and scalable data collection. Due to the unique requirements of sensing tasks and the mobility of users, a service provider should allocate the sensing tasks to the proper mobile users who are/will be in the sensing area. For instance, to measure the traffic congestion in downtown, Toronto, the service provider should recruit the mobile users driving on the roads in downtown, Toronto. Fog computing can achieve local management of mobile users and capture the mobility patterns of mobile users in its coverage area, and thereby assist the service provider to achieve accurate task allocation. Specifically, the service provider will first assigns the sensing tasks to the fogs physically located in the intended sensing area, and then the fogs, acting as geography-related local servers, further finds proper mobile users to perform the tasks. This two-step task allocation approach not only decreases the cost of the service provider, but also improves the accuracy of task allocation.

1.2.2.3 Smart Grid

In smart grids, the operation center frequently collects the real-time power consumption of every household using smart meters, which is utilized to optimize energy generation, power distribution and electricity billing. The smart meters, deployed in houses, independently report power usage measurements to the operation center, which brings unacceptable communication overhead between smart meters and the operation center. On behalf of fogs, Advanced Metering Regional Collectors (AMRC) are deployed to the public to aggregate the consumption measurements and submit the sum of power consumption in their residential areas to the operation center [16]. In this way, the burden on communication bandwidth between the operation center and AMRC is dramatically decreased. Besides, AMRC can also transiently store the individual power usage reports and calculate the daily cost on power for each household based on dynamic electricity price, such that real-time load monitoring and dynamic billing can be realized, simultaneously.

1.3 Security and Privacy Challenges

In data-intensive IoT, fogs utilize their resources to perform the tasks of data pre-processing for latency reduction and network throughput improvement. It is recognized that fog computing is a more secure paradigm than cloud computing due to the property of localization, but the security threats faced by fog computing are equally as serious as cloud-based IoT environments [17]. For example, due to the limited computing and storage resources, a fog is vulnerable to the distributed denial-of-service (DDoS) attack. In addition, mobile devices have become new weapons for hackers, including mobile Botnets, ransomware and IoT malware. These threats introduce security concerns towards users, as well as creating serious security vulnerabilities to the IoT applications. As a result, it leads to data corruption, data leakage, or application deprecation [18].

User privacy may be exposed during data processing and storage in fog-enabled IoT. The fogs extract users' personal information from the data submitted by the devices [19]. We discuss three types of user privacy leakage: identity privacy leakage, location privacy leakage, and data privacy leakage.

Identity Privacy Leakage The identity information of a user includes name, identity number, address, telephone number, social insurance number, license number, visa number and public-key certificate in which any of the information listed may link to a specific user. The identity of a user is vulnerable to be leaked out from different data submitted to fogs for authentication.

Location Privacy Leakage Thanks to the location awareness of base stations and Wifi hotspots, a device's position can be extracted based on the access points. The limited coverage of fog increases the threat of location privacy exposure.

Even though it is a coarse-grained location exposure, the activity area of a specific device owner is exposed from the public location. Moreover, if a device has connections with multiple fogs, the precise location can be obtained using positioning techniques.

Data Privacy Leakage In IoT services, the collected data encapsulate different aspects of physical environment, some may be considered sensitive, such as personal activities, preferences, health status and industrial design drawing, but other may not, such as air pollution index, social events and public information. It is widely believed that the data ownership should belong to data owners. Nevertheless, to explore data utility, data are frequently shared with others without the permission of their owners. Consequently, the data, either sensitive or not, would be captured or accessed by the ineligible entities.

1.4 Aim of the Monograph

The aim of this monograph is to investigate how to preserve user privacy in fog-enabled IoT applications by utilizing privacy-enhancing technologies. Firstly, we show a literature review on the state-of-the-art privacy-enhancing technologies, which achieve identity privacy, location privacy and data privacy protection. Then, we propose three privacy-preserving schemes to address the identity leakage, location leakage and data leakage in smart parking, mobile crowdsensing and smart grid, respectively.

In Chap. 3, we investigate the identity information leakage for drivers, when they are accessing the smart parking systems [20]. To acquire needed parking information, the drivers report personal queries to access parking spaces in the desirable destinations, which ends up with identity privacy violation if the queries are not protected. To preserve drivers' privacy, we propose a privacy-preserving smart parking navigation scheme (P-SPAN) with efficient navigation result retrieval for drivers using Bloom filters. P-SPAN allows a cloud and multiple fogs to guide vehicles to vacant parking spaces in the destinations based on real-time parking information without disclosing any personal information about drivers. In addition, an efficient data retrieval mechanism is developed to achieve navigation result retrieval for querying vehicles. The drivers can anonymously query available parking spots to the cloud, and efficiently retrieve the encrypted navigation results from the passing-by fogs. P-SPAN can provide identity privacy-preserving parking navigation with high retrieving probability on navigation results and low computational and communication overhead.

In Chap. 4, we study the location information leakage for mobile users when they are participating in mobile crowdsensing activities [21]. In mobile crowdsensing, to enhance data trustworthiness, it is critical for service provider to recruit mobile users based on their personal features, e.g., mobility pattern and reputation, but it leads to the privacy leakage of mobile users. To address location privacy leakage, we propose a fog-assisted privacy-preserving mobile crowdsensing framework that enables fogs to allocate tasks based on user mobility for improving the accuracy of task assignment. Specifically, in the first step, the service provider allocates the sensing tasks to fogs based on the locations of fogs and the sensing areas of tasks; and in the second step, the fogs can recruit mobile uses in this coverage areas without knowing any knowledge about the locations and trajectories of mobile users. Therefore, the location of mobile users is protected during task allocation in mobile crowdsensing.

In Chap. 5, we focus on the data privacy leakage for customers when they are submitting the real-time power consumption to the operation center through collectors in smart grid [22]. Real-time power consumption brings serious privacy issues to customers, since the meter readings can possibly disclose customers' activities in the house. Although secure data aggregation improves communication efficiency and preserves customers' privacy, it fails to support dynamic billing, or guarantee integrity protection against public fogs. Therefore, we define a new security model to formalize the misbehavior of fogs, in which the misbehaving fogs may launch pollution attacks to corrupt power consumption data, and propose a privacy-preserving smart metering scheme to prevent pollution attacks for the balance of privacy and efficiency. The proposed scheme supports end-to-end security, data aggregation and integrity protection against the misbehaving fogs. The proposed scheme achieves secure smart metering and verifiable dynamic billing against misbehaving fogs with low computational and communication overhead.

Finally, we discuss the future research directions on fog computing, including how to detect rogue fogs and IoT devices, how to prevent privacy exposure in data combination and how to build decentralized and scalable secure infrastructure, and demonstrate our insights to reach these research directions for building secure and privacy-preserving fog-enabled IoT applications.

References

1. A. Al-Fuqaha, M. Guizani, M. Mohammadi, M. Aledhari, and M. Ayyash, "Internet of things: A survey on enabling technologies, protocols, and applications," *IEEE Communications Surveys & Tutorials*, vol. 17, no. 4, pp. 2347–2376, 2015.
2. A. Botta, W. De Donato, V. Persico, and A. Pescapé, "Integration of cloud computing and internet of things: a survey," *Future Generation Computer Systems*, vol. 56, pp. 684–700, 2016.
3. C.-W. Tsai, C.-F. Lai, M.-C. Chiang, L. T. Yang *et al.*, "Data mining for internet of things: A survey." *IEEE Communications Surveys and Tutorials*, vol. 16, no. 1, pp. 77–97, 2014.
4. J. Ni, Y. Yu, Y. Mu, and Q. Xia, "On the security of an efficient dynamic auditing protocol in cloud storage," *IEEE Transactions on Parallel and Distributed Systems*, vol. 25, no. 10, pp. 2760–2761, 2014.
5. F. Jalali, K. Hinton, R. Ayre, T. Alpcan, and R. S. Tucker, "Fog computing may help to save energy in cloud computing," *IEEE Journal on Selected Areas in Communications*, vol. 34, no. 5, pp. 1728–1739, 2016.
6. F. Bonomi, R. Milito, J. Zhu, and S. Addepalli, "Fog computing and its role in the internet of things," in *Proc. of MCC*, 2012, pp. 13–16.

7. M. Chiang, S. Ha, I. Chih-Lin, F. Risso, and T. Zhang, "Clarifying fog computing and networking: 10 questions and answers," *IEEE Communications Magazine*, vol. 55, no. 4, pp. 18–20, 2017.
8. P. Garcia Lopez, A. Montresor, D. Epema, A. Datta, T. Higashino, A. Iamnitchi, M. Barcellos, P. Felber, and E. Riviere, "Edge-centric computing: Vision and challenges," *ACM SIGCOMM Computer Communication Review*, vol. 45, no. 5, pp. 37–42, 2015.
9. J. Ni, K. Zhang, Y. Yu, X. Lin, and X. Shen, "Providing task allocation and secure deduplication for mobile crowdsensing via fog computing," *IEEE Transactions on Dependable and Secure Computing*, 2018.
10. P. Hu, H. Ning, T. Qiu, Y. Zhang, and X. Luo, "Fog computing based face identification and resolution scheme in internet of things," *IEEE transactions on industrial informatics*, vol. 13, no. 4, pp. 1910–1920, 2017.
11. P. Garcia Lopez, A. Montresor, D. Epema, A. Datta, T. Higashino, A. Iamnitchi, M. Barcellos, P. Felber, and E. Riviere, "Edge-centric computing: Vision and challenges," *ACM SIGCOMM Computer Communication Review*, vol. 45, no. 5, pp. 37–42, 2015.
12. M. Chiang and T. Zhang, "Fog and iot: An overview of research opportunities," *IEEE Internet of Things Journal*, vol. 3, no. 6, pp. 854–864, 2016.
13. M. Tao, K. Ota, and M. Dong, "Foud: integrating fog and cloud for 5g-enabled v2g networks," *IEEE Network*, vol. 31, no. 2, pp. 8–13, 2017.
14. J. Ni, K. Zhang, X. Lin, Y. Yu, and X. Shen, "Cloud-based privacy-preserving parking navigation through vehicular communications," in *Proc. of Securecomm*, 2016, pp. 85–103.
15. J. Ni, A. Zhang, X. Lin, and X. Shen, "Security, privacy, and fairness in fog-based vehicular crowdsensing," *IEEE Communications Magazine*, vol. 55, no. 6, pp. 146–152, 2017.
16. J. Ni, K. Zhang, K. Alharbi, X. Lin, N. Zhang, and X. Shen, "Differentially private smart metering with fault tolerance and range-based filtering," *IEEE Transactions on Smart Grid*, vol. 8, no. 5, pp. 2483–2493, 2017.
17. J. Ni, K. Zhang, X. Lin, and X. Shen, "Securing fog computing for internet of things applications: Challenges and solutions," *IEEE Communications Surveys & Tutorials*, vol. 20, no. 1, pp. 601–628, 2017.
18. J. Granjal, E. Monteiro, and J. S. Silva, "Security for the internet of things: a survey of existing protocols and open research issues," *IEEE Communications Surveys & Tutorials*, vol. 17, no. 3, pp. 1294–1312, 2015.
19. S. J. Stolfo, M. B. Salem, and A. D. Keromytis, "Fog computing: Mitigating insider data theft attacks in the cloud," in *Proc. of SPW*, 2012, pp. 125–128.
20. J. Ni, K. Zhang, Y. Yu, X. Lin, and X. Shen, "Privacy-preserving smart parking navigation supporting efficient driving guidance retrieval," *IEEE Transactions on Vehicular Technology*, vol. 67, no. 7, pp. 6504–6517, 2018.
21. J. Ni, K. Zhang, Q. Xia, X. Lin, and X. Shen, "Enabling strong privacy preservation and accurate task allocation for mobile crowdsensing," *arXiv preprint arXiv:1806.04057*, 2018.
22. J. Ni, K. Zhang, X. Lin, and X. Shen, "Balancing security and efficiency for smart metering against misbehaving collectors," *IEEE Transactions on Smart Grid*, 2017.

Chapter 2
Privacy-Enhancing Technologies

In this chapter, we will review several state-of-the-art privacy-enhancing techniques for identity, location and data privacy preservation.

2.1 Identity Privacy-Enhancing Techniques

We review several identity privacy protection techniques, including pseudonymization, k-anonymity, mix-network, blind signature, group signature and ring signature.

2.1.1 Pseudonymization

The most popular way to protect user identity online is to use pseudonyms. Instead of using a real identity, people assume one or multiple artificial identifiers, or pseudonyms. It is called pseudonymization. In [1], Raya and Hubaux introduced a concept of anonymous certificate (or anonymous key pairs) for secure communication and privacy protection in Vehicular ad hoc networks (VANETs), shown in Fig. 2.1. In traditional Public key infrastructure (PKI), a digital certificate is issued by a trusted authority, known as Certificate Authority (CA). It is comprised of a user's identity and his/her public key, and most importantly, the binding of the public key and the identity is assured by a CA's digital signature. In other words, a digital certificate proves the ownership of a public key. In an anonymous certificate, the public key is also authenticated by a CA through its signature on the key, but unlike traditional public key certificate, there is no actual identity information within the certificate. Generally speaking, you can treat anonymous certificates as pseudonyms. When a vehicle sends out a message, it signs the message using the private key corresponding to an anonymous certificate and then broadcasts it along with

X. Lin et al., *Privacy-Enhancing Fog Computing and Its Applications*,
SpringerBriefs in Electrical and Computer Engineering,
https://doi.org/10.1007/978-3-030-02113-9_2

the anonymous certificate. When the message arrives at another vehicle, the receiver checks the certificate's validity by using the CA's public key, and then verifies the sender's signature on the message with the public key from the certificate. By doing so, the receiver can be assured that the message received comes from a legitimate vehicle. However, the receiver does not know the real identity of the sender. Thus, the users' identity privacy in VANETs is protected. Note that identity privacy does not guarantee location privacy. One appealing solution to achieve location privacy in VANETs is that vehicles periodically change their pseudonyms when broadcasting messages to other vehicles. When a vehicle use multiple different pseudonyms on the road, an observer cannot link multiple pseudonyms from a vehicle together. The unlinkability of pseudonyms can guarantee a vehicle's location privacy. Thus, it is very common to have a vehicle possess a large set of anonymous certificates in order to achieve both identity privacy and location privacy.

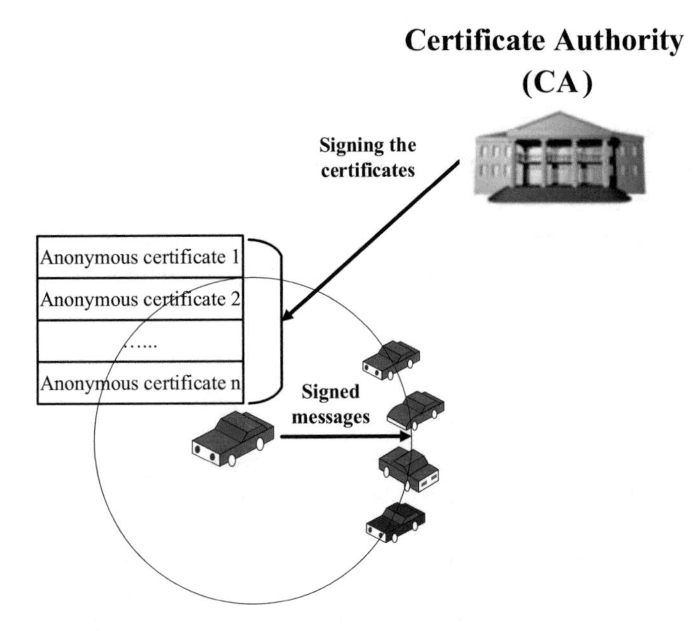

Fig. 2.1 Secure communication and privacy protection in VANETs

2.1.2 *k-Anonymity*

k-anonymity is an anonymous property that protects the privacy of released dataset against the re-identification attack such that any data provider, to whom the released data refer, is indistinguishable from no less than k other data providers.

Anonymizing the data is a basic solution for achieving data privacy protection, but how to achieve the anonymity property is an open problem for many years.

Among all anonymous properties, k-anonymity is the most famous one that was first proposed by Samarati and Sweeney [2]. Sweeney then discussed this concept in-depth and showed its relationship to data privacy protection in his following works [3, 4].

The idea of k-anonymity comes from the observation that publicly released dataset, though explicit identities of data providers (e.g., name and phone number) are concealed or removed, may still contain some data attributes such as gender and birth date, which can be utilized for identifying a unique and specific individual. These data attributes are called as quasi-identifiers. Specifically, quasi-identifiers are pieces of information that are not of themselves unique identifiers, but they are correlated with each other and can be combined together to create a unique identifier for any data provider and re-identify the anonymous dataset. To protect the publicly released dataset from such correlation attack and re-identification attack, k-anonymity requires that each distinct quasi-identifier should belong to no less than k individuals unconditionally. To achieve this purpose, two techniques, named generalization and suppression, were proposed at a microdata level. Generalization means any quasi-identifier, e.g., detailed birth date, is replaced by a more general value, e.g., year of birth date. The result after the generalization operations shows that quasi-identifiers with different values are substituted by the same value, thus becoming indistinguishable. Suppression, similarly, can be viewed as a stronger version of generalization. Suppression means that each quasi-identifier, e.g., gender: male and female, is replaced by the asterisk '$*$', thus becoming indistinguishable since they are same. In general, generalization and suppression can be applied at different levels of granularity according to different privacy requirements.

While k-anonymity is a promising anonymous approach given by its simplicity, it may suffer from homogeneity attacks and background knowledge attacks [5]. k-anonymity only considers the protection of explicit attributes and quasi-identifiers but ignores that sensitive attributes, e.g., disease, may also leak the privacy of data provider. Hence, the concept of l-diversity was proposed by Machanavajjhala et al. [5] that was built based on k-anonymity to ensure that sensitive attributes must be diverse within each quasi-identifier equivalence class. Nevertheless, l-diversity is also not perfect since it is neither necessary nor sufficient under some conditions. For example, HIV records are sensitive attributes but only include two values: positive and negative. These two values have very different degrees of sensitivity. One would not mind being known to be tested negative, because it indicates that he/she is the same as 99% of the population, but one would not want to be considered as positive. In this case, 2-diversity is unnecessary for an equivalence class that contains only records that are negative. In addition, l-diversity does not consider the real-world distribution of sensitive attributes and the semantics of sensitive attributes, which may cause privacy risks to arise. Thus, the concept of t-closeness was proposed by Li et al. [6] that formally defined the distance between the distribution of a sensitive attribute in an equivalence class and the distribution of the attribute in the whole dataset. t-closeness requires that the distance is no less than a threshold t, so the l-diversity can become more realistic.

k-anonymity and its extensions, like l-diversity and t-closeness, are still being investigated in recent years and there are many related works presented in this area.

Usecase of k-anonymity A typical usecase of k-anonymity is to protect the patient's privacy related to his/her medical records. As shown in Table 2.1, the public dataset of various patients are stored with different attributes (i.e. key attribute: name; quasi-identifiers: birth date, gender and zip code; sensitive attribute: disease).

Table 2.1 Medical records of patients

Name	Birth date	Gender	Zip code	Disease
Alice	01/01/1991	Female	N2L3W9	Heart Disease
Bob	02/02/1992	Male	N2L2N5	Hepatitis
Carl	03/03/1993	Male	N2L2Q3	Brochitis
Dan	04/04/1991	Male	N2L1X7	Broken Arm
Ellen	05/05/1992	Female	N2L2P0	Flu
Fazio	06/06/1993	Male	N2L3J4	Hang Nail

To achieve k-anonymity ($k = 6$), the key attribute (name) should be removed and the quasi-identifiers (birth date, gender and zip code) should be generalized or suppressed. The final result is shown in Table 2.2. In this situation, even if the attacker knows that Alice (Female) living at N2L3W9, who was born in 01/01/1991, exist among the records, he/she cannot distinguish Alice's sensitive disease from the whole six kinds of diseases. Therefore, Alice's privacy is protected under k-anonymity ($k = 6$).

Table 2.2 Anonymous medical records of patients

Name	Birth date	Gender	Zip code	Disease
	$\leq$1994	$*$	N2L***	Heart Disease
	$\leq$1994	$*$	N2L***	Hepatitis
	$\leq$1994	$*$	N2L***	Brochitis
	$\leq$1994	$*$	N2L***	Broken Arm
	$\leq$1994	$*$	N2L***	Flu
	$\leq$1994	$*$	N2L***	Hang Nail

2.1.3 Mix-Network

Mix networks are routing protocols that achieve hard-to-trace communications in the public network, in order to protect the privacy of senders and receivers. A bunch of proxy servers will form a chain of mix-net servers (a.k.a mixers) and each of them will receive the messages from multiple senders or previous mixers, shuffles them randomly and sends them to the next mixers or destination such that the link between the source of the message and the destination is broken.

Untraceability is an indispensable property for communication privacy over public networks. To achieve untraceability, Chaum [7] first proposed the concept of mix-networks (a.k.a mixnets) in 1981, which provides sender and receiver anonymity, i.e., the originator and the recipient of a message are difficult to be discerned from the perspective of attackers. Generally speaking, there mainly exist two types of mix-networks: decryption mixnets and re-encryption mixnets.

The decryption mixnets were initiated in [7], where the sender is required to encrypt the message with every keys of mixers. When the message goes through each mixer, the mixer uses its corresponding key to decrypt the message, shuffles the message with other received messages and delivers them to the next mixer until the receiver receives the message. This category of mixnets can support both symmetric cryptosystems, e.g., AES, and public key cryptosystems, e.g., RSA and Elgamal. Compared to the public key cryptosystems, symmetric cryptosystems are more efficient in decryption mixnets but the sender needs to share the corresponding symmetric key with each mixer in advance.

The re-encryption mixnets [8] were built based on the proxy re-encryption technique, where the ciphertext under one public key can be transferred into the ciphertext under a different public key, using a re-encryption key. Namely, the sender only encrypts the message with the first mixer's public key and submits the message to that mixer. The mixer then re-encrypts the message to the next mixer's ciphertext, shuffles the message with other messages and delivers them to the next mixers until the receiver receives the message. Obviously, this category of mixnets can only support the public key cryptosystem due to the re-encryption operation, such as Elgamal. To further develop the re-encryption mixnets, the universal re-encryption mixnets were proposed by Golle et al. [9], which does not need the mixers to hold the public key and private key pairs and reduce the costs of cumbersome requirements such as key generation, key distribution, and private-key management.

When deploying the mixnets, one of the important issues is how to check the correctness of the mixnets since not all mixers are honest and the mixers could be compromised and behave maliciously. The correctness of mixnets can be analyzed based on three criteria: (1) the input messages from the previous mixers have been transformed (decrypted or re-encrypted) and permuted honestly; (2) the input messages in the input batch have not been corrupted; (3) inputs have not been added/deleted in the mixnets. Hence, the concept of verifiable mixnets (shuffle) [10] have been proposed based on the zero-knowledge proof technique, which provides verifiability property for the mixnets. However, the verifiable mixnets can only be designed based on the public key cryptosystem, which is time-consuming. How to improve the efficiency is another issue waiting to be solved. Recently, Pereira and Rivest [11] proposed an efficient mixnets, called marked mixnets, which can achieve additional assurances about the privacy of the messages, compared to non-verifiable mixnets.

Mixnets are still being in development in terms of different requirements of real-world applications. Except mixnets, there also exist some similar ideas such as onion routing to preserve the anonymity of users.

Usecase of Mix-Networks A typical usecase of mix-networks is the electronic secret ballot (a.k.a E-voting), to protect the voter's privacy in a way such that the adversary has no knowledge about the voter's choice. In this case, a simple voting system is presented with four voters (V_1, V_2, V_3 and V_4), two candidates (C_1, C_2) and two mixers (MIX_1 and MIX_2). The public keys of two mixers are PK_1 and PK_2, and the public key encryption with randomness can be represented as $E(M, PK)$, where M is the message and PK is the public key.

As shown in Fig. 2.2, each voter first encrypts its choice ($V_1 \rightarrow C_1$,$V_2 \rightarrow C_1$,$V_3 \rightarrow C_2$ and $V_4 \rightarrow C_2$) using the public keys of two mixers as the ciphertext $ct = E(E(C_?, PK_2), PK_1)$. Then, each voter submits the choice to the first mixer MIX_1. After the mixer decrypts each voter's ciphertext and permutes the results, it sends the results to the second mixer MIX_2. Following the same operations, it can finally reveal the voting results (2 votes for C_1 and 2 votes for C_2) without exposing the voter's privacy.

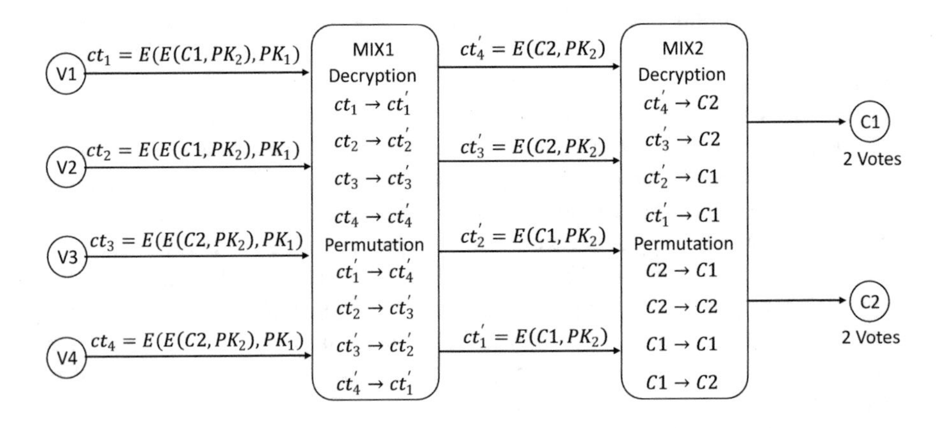

Fig. 2.2 Mix-network for e-voting

2.1.4 Blind Signature

A blind signature [12] scheme is an interactive protocol between a user and a signer. It allows the user to obtain a signature on a message form the signer without leaking the message or the signature to the signer. An ID-based blind signature is scheme is an integration of an identity-based signature scheme [13] and a blind signature scheme. Specifically, it consists of five tuples:

- **TA** is a trusted authority which is charge of system setup and user private key generation.
- **Setup** is run by **TA** that takes a security parameter k and outputs system parameters $para$.

- **Extract** takes as input $para$ and a random $ID \in \{0, 1\}*$ and returns a private key S_{ID}.
- **Sign** is an interactive protocol executed by a signer and a user. User takes as inputs a signer's identity and a message (ID, m). Signer takes as inputs its identity and private key (ID, S_{ID}). At the end of the protocol, signer outputs *complete* or *not-complete*, and the user outputs *failed* or the signature $\sigma(m)$ on the message m.
- **Verify** takes as inputs $(ID, \sigma(m), m)$ and outputs either *accept* or *reject*.

An example of the identity-based signature scheme is described below.

Setup Let $\mathbb{G}$ be a group with prime order q, a generator P, and a bilinear pairing: $e : G \times G \rightarrow V$. **TA** defines two collision-resistant hash functions $H : \{0, 1\}* \rightarrow \mathbb{Z}_q$ and $H_1 : \{0, 1\}* \rightarrow \mathbb{G}$. **TA** chooses a secret $s \in_R \mathbb{Z}_q$ and computes $P_{pub} = sP$. The public parameters $para$ of the system are:

$$para = (\mathbb{G}, q, P, P_{pub}, H, H_1) \tag{2.1}$$

Extract Given an identity ID, the public key of the ID is $Q_{ID} = H_1(ID)$. **TA** can compute a private key $S_{ID} = sQ_{ID}$.

Sign The signer generates the signature on the message.

1. The signer chooses a random number $r \in \mathbb{Z}_q$ and computes $R = rP$. The signer sends R to the user.
2. The user chooses two random numbers $a, b \in \mathbb{Z}_q$, known as blind factor, and computes

$$t = e(bQ_{ID} + R + aP, P_{pub}), c = H(m, t) + b(modq) \tag{2.2}$$

   The user sends c to the signer.
3. The signer computes

$$S = cS_{ID} + rP_{pub} \tag{2.3}$$

   and sends S to the user.
4. The user computes

$$S' = S + aP_{pub}, c' = cb \tag{2.4}$$

   and outputs (S', c') as the signature σ on the message m.

Verify The signature can be accepted if

$$c' \stackrel{?}{=} H(m, e(S', P)e(Q_{ID}, P_pub)c') \tag{2.5}$$

Please note that the final signature (S', c') is a different version than S which was generated by the signer. Therefore, the two signatures (S', c') and S cannot be linked and as a results, preserves the anonymity if the signature (S', c') is used later.

One promising use case for blind signatures is cryptocurrency. Specifically, we can use the blind signature technique to create a digital analogue of cash that can protect the privacy and anonymity of the user. A scenario is described as follows, shown in Fig. 2.3

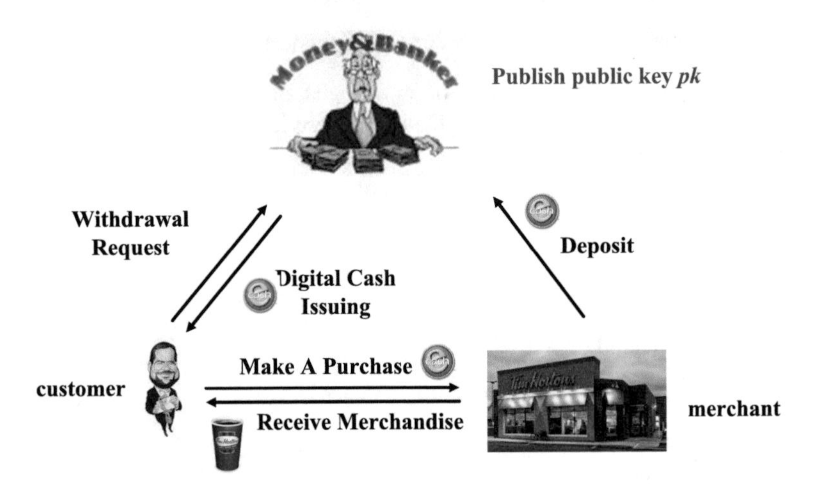

Fig. 2.3 Digital cash scenario using blind signature

Suppose that a bank publishes his public key pk, and keeps the corresponding private key sk confidential. For simplicity, if the bank uses this private key sk to sign a digital note, it means the signed note represents a certain amount of money, such as one-dollar bill. Then, it can be used and spent once, as if it was the same as the bill, to make a purchase from a merchant, for this example we will be using Tim Hortons as our merchant. When a bank customer makes a withdrawal from his or her account, the bank provides the customer with a digitally signed note using the blind signature technique. Note that the bank cannot link the note to the customer who made a cash withdrawal since blind signatures have been used. Afterwards, the customer can present it to Tim Hortons, which can then verify the bank's signature on the note (digital cash) using the public key of the bank. Upon completing a transaction, Tim Hortons can then remit the note to the bank, which will then credit Tim Hortons one dollar.

2.1.5 *Group Signature*

A group signature scheme can provide anonymity for signers in a group. Each member has its own private signing key, by using it the group member can produce signatures on behalf of the group. The signatures keep the identity of the signer secret. But there is an authority who owns an opening key can trace the signature and identify the user who signed the signature. A group signature consist of six algorithms [14]:

- **GSetup** is a setup algorithm that takes a security parameter k as input and output a group public key gpk and the group manager's secret key $gmsk$.
- **PKIJoin** is run by a user who wants to join the group. It takes as input the security parameter k and the index of the user i. The output of the algorithm is a personal public and private key pair $(upk[i], usk[i])$. The upk is public and anyone can get an authentic copy of it.
- **GJoin** is an interactive protocol between the user and the group manager and is used to add new users to the group. The input of the user is its index i, the secret key $usk[i]$ and the group public key gpk. The input of the group manager is the user index i, the user public key $upk[i]$ and the group manager's secret key $gmsk$. At the end of the protocol, if the user joins the group successfully, the group manager will add the index i in its registration table reg. The user will get its private signing key $gsk[i]$.
- **GSign** allows a group member to sign on behalf of the group and obtain a signature on a message m. it takes as input a users private signing key $gsk[i]$ and a message m, and outputs a signature σ.
- **GVerify** can be used to verify the signature. It takes as input the group public key gpk, a message m and a group signature σ of m. The signature is valid if the output is 1.
- **GOpenn** is run by the data owner to open a signature. It can output the identity of a group member who produced the signature. It takes as input the group manager's secret key $gmsk$, a message m, a signature of the message σ and the registration table reg, and outputs the a user index i and a proof π which can used to verify the user i created the signature. The algorithm returns $\perp$ if the opening fails.

An example of the group signature scheme is described below [15].

GSetup(1^k) Given a security parameter k, the algorithm output the public parameter $pp = (p, G_1, G_2, G_T, e)$. Then the group manager selects $\widetilde{g} \leftarrow G_2$, $(x, y) \leftarrow Z_p^2$, and computes $(\widetilde{X}, \widetilde{Y}) \leftarrow (\widetilde{g}^x, \widetilde{g}^y)$. The group managers secret key $gmsk = (x, y)$, the group public key $gpk = (\widetilde{g}, \widetilde{X}, \widetilde{Y})$.

PKIJoin$(i, 1^k)$ Let Σ be a digital signature scheme, the user generates $(usk[i], upk[i]) \leftarrow \Sigma.Keygen(1^k)$ and the $upk[i]$ is public.

GJoin It is an interactive protocol between a user and the group manager. The user first generates a secret $sk_i \leftarrow Z_p$, and sends $(\tau, \widetilde{\tau}) \leftarrow (g^{sk_i}, \widetilde{Y}^{sk_i})$ and the signature $\eta \leftarrow \Sigma.Sign(usk[i], \tau)$ to the group manager. The group manager verifies the validity of the η and check $(\tau, \widetilde{\tau})$ by testing whether $e(\tau, \widetilde{Y} = e(g, \widetilde{\tau}))$. Then, the user starts an interactive proof of knowledge of sk_i. If the proof succeeds, the group manager chooses $u \leftarrow Z_p$ randomly and computes $\sigma \leftarrow (\sigma_1, \sigma_2) \leftarrow (g^u, (g^x, (\tau)^y)^u)$. The group manager stores $(i, \tau, \eta, \widetilde{\tau})$ in the registration table and sends σ to the user. The private signing key of the user is $gsk_i = (sk_i, \sigma, e(\sigma_1, \widetilde{Y}))$.

GSign(gsk_i, m) To generate a signature on a message m, the user first generates a random t and computes $(\sigma_1^{'}, \sigma_2^{'}) \leftarrow (\sigma_1^t, \sigma_2^t)$. Then, to generate a signature of knowledge of sk_i, the user chooses a random $k \leftarrow Z_p$ and computes $e(\sigma_1^{'}, \widetilde{Y})^k \leftarrow e(\sigma_1, \widetilde{Y})^{kt}$, $c \leftarrow H(\sigma_1^{'}, \sigma_2^{'}, e(\sigma_1, \widetilde{Y})^{kt}, m)$ where H is a hash function. Finally, the user computes $s \leftarrow k + c \cdot sk_i$ and the group signature μ is $(\sigma_1^{'}, \sigma_2^{'}, c, s) \in G_1^2 \times Z_p^2$.

GVerify(gpk_i, m) Given a signature $\mu = (\sigma_1, \sigma_2, c, s)$ of m, the verifier computes $R \leftarrow (e(\sigma_1^{-1}, \widetilde{X}) \cdot e(\sigma_2, \widetilde{g}))^{-c} \cdot e(\sigma_1^s, \widetilde{Y})$ and verifies if $c = H(\sigma_1, \sigma_2, R, m)$. If the equation holds, the verifier outputs 1, otherwise outputs 0.

GOpen($gmsk, m, \mu$) Given a signature μ which needs to be opened, the group manager check all the entries $(i, \tau_i, \eta_i, \widetilde{\tau_i})$ whether $e(\sigma_2, \widetilde{g}) \cdot e(\sigma_1, \widetilde{X})^{-1} = e(\sigma_1, \widetilde{\tau})$. If the match succeeds, the group manager outputs the corresponding entry (i, τ_i, η_i) and a proof of the knowledge of $\widetilde{\tau_i}$ which can be used to check the correctness of the opening.

Group signature schemes can be used in the vehicular ad hoc network (VANET). In the intervehicle communication, we can use the group signatures to guarantee the anonymity of the users and when there is a traffic event dispute, the authority can reveal the identity of the message sender. The scenario is described as follows.

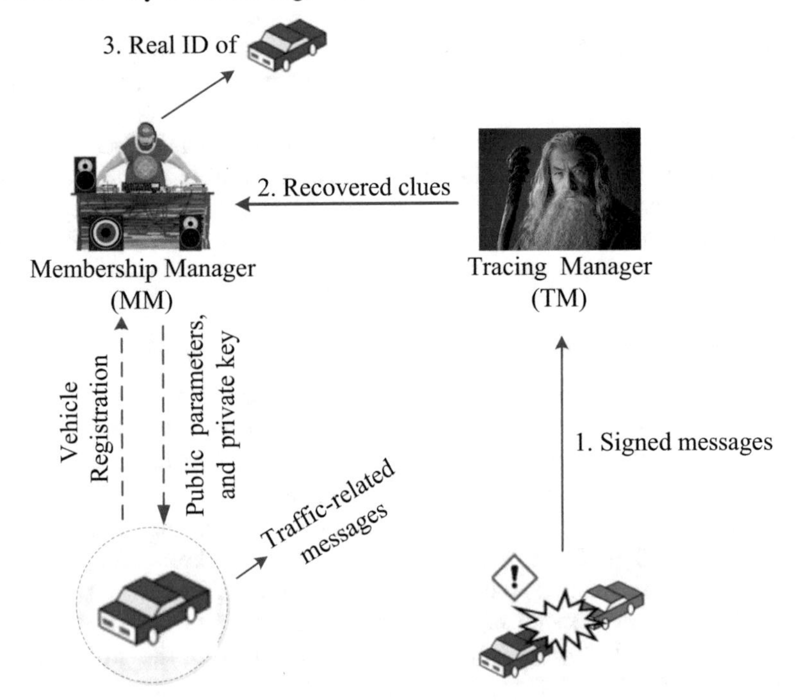

Fig. 2.4 VENET scenario using group signature

As shown in Fig. 2.4, suppose that with the communication devices equipped in the vehicles, they can communicate with each other. A group signature scheme is used to sign the messages sent by the vehicles [16]. The Membership manager

(MM) is responsible for assigning group public keys and private signing keys to the vehicles. Vehicles need to be registered with the MM and preloaded with the private keys and public system parameters before they join the VANET. Vehicles can send traffic-related messages to each other and signed with their private keys. Receivers of the messages do not know the identity of the signer. However, When the ID of a vehicle need to be revealed, for example, someone can provide valuable information about an accident, its signed messages can be submitted to the Tracing Manager (TM) who is responsible for revealing the real identity of the specific vehicles. Then the TM send the recovered clues to the MM who will finally find the real identity from its registration database.

2.1.6 Ring Signature

A ring signature is a simplified group signature which consists of only users without managers. It provides the anonymity of the signer which allows the verifier to know that the signature is signed by a member of the ring, but can not know who the signer is. An ID-based ring signature can be viewed as a combination of an ID-based signature and a ring signature. Specifically, An ID-based ring signature consists of four tuples [12].

- **Setup** is run by TA that takes as input a security parameter k and outputs the public parameters PP and the master key s.
- **Extract** is run by TA. It takes as input the public parameters, the master key, and an arbitrary $ID \in \{0, 1\}^*$ which is the signers identity and can used as the public key, and outputs a private key S_{ID}.
- **Signing** takes as input the public parameters, a private key S_{ID}, a list of identities L and a message m and outputs a signature $\sigma(m)$ of m.
- **Verification** takes $(L, m, \sigma(m))$ as input and outputs either accept or reject.

An example of the identity-based ring signature scheme is described as follows [12].

Setup Let P is a generator of a group G. TA chooses a random $s \in Z_q^*$ and set $P_{pub} = sP$. TA then define two hash functions $H : \{0, 1\}^* \rightarrow Z/q$ and $H_1 : \{0, 1\}* \rightarrow G$. The public parameters are $PP = \{G, q, P, P_{pub}, H, H_1\}$. The master key of TA is s.

Extract Let $L = \{ID_i\}$ be the set of identities. Given an identity ID, the public key of the ID is $Q_{ID} = H_1(ID)$. TA computes $S_{ID} = sH_1(ID)$ as the private key associated with ID.

Signing The signer generate a signature of a message as follows.

1. The signer chooses a random element $A \in G$ and computes $c_{k+1} = H(L \parallel m \parallel e(A, P))$.
2. For $i = k + 1, \cdots, n - 1, 0, 1, \cdots, k - 1$, the signer chooses randomly $T_i \in G$ and computes $c_{i+1} = H(L \parallel m \parallel e(T_i, P)e(c_i H_i(ID_i), P_{pub}))$.

3. The signer computes $T_k = A - c_k S_{ID_k}$.
4. The signer selects 0 as the glue value, and outputs the $(n + 1)$-tuple: $c_0, T_0, T_1, \cdots, T_{n-1}$ as the signature for m and L.

Verification Given $c_0, T_0, T_1, \cdots, T_{n-1}, m$, and L, the verifier computes $c_{i+1} = H(L \parallel m \parallel e(T_i, p)\, e(c_i H_1(ID_i), P_{pub}))$, for $i = 0, 1, \cdots, n-1$.

If $c_n = c_0$, the signature is accepted, otherwise, it is rejected.

Ring signature can be used in the electronic auction where the anonymity of the bidders needs to be guaranteed. The ring signature technique allows auctioneer to determine the winning bid without revealing the losing bid. A scenario is described as follows and the auction procedure based on ring signature is shown in Fig. 2.5.

A sealed-bid electronic auction [17] consists of two phases: the bidding period in which bidders can choose a bid from a set of biddable values, and sends the sealed bids to the auction manager (AM). When the bidding period is finished, the opening period allow the AM open the bid which is the highest and determine the winner. In the bidding period, the bidder sends the registration information to the auction manager who holds the auction and manages a Bulletin Board System (BBS). Then, the bidder chooses bid he wants to pay, computes the ring signature of the bid, and sends them to the AM through an anonymous connection. In the opening phase, by some calculation, the AM can find the winner who cast the highest price. If the bidder cheats or crashes, with the help of the T who is a trust third party and is responsible for resolving a dispute. After the malicious bidder is revoked by T, AM can exclude the malicious bidder from bidder's list in the BBS and continue the opening bids.

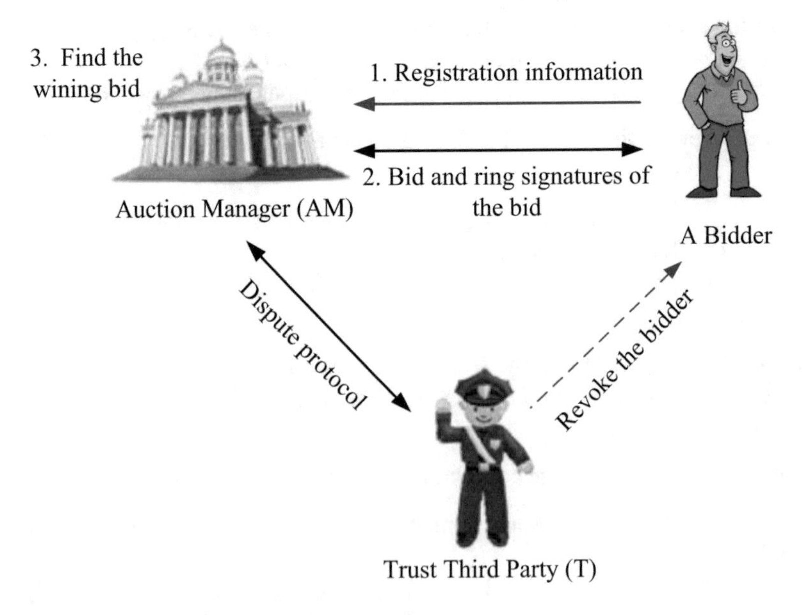

Fig. 2.5 Electronic auction scenario using ring signature

2.2 Location Privacy-Enhancing Techniques

We review several location privacy protection techniques, including obfuscation, differential location privacy and privacy-preserving location matching.

2.2.1 *Obfuscation*

The physical position is sensitive information for users. Assume that positions of users can be denoted as planar circular areas. A location measurement can be defined as follows:

Location measurement [18] Let (x_u, y_u) be the real position of a user. The location measurement is a circular area $A_i =< x_i, y_i, A_i >$, such that (x_i, y_i) are the coordinates of the center of A_i, r_i is the radius, and the following conditions are satisfied.

- $P((x_u, y_u) \in A_i) = 1$;
- $P((x_u, y_u) \in A_i)$, where $A =< x, y, \eta r > \subset A_i$ is the neighborhood of position (x, y) with ηr an infinitely small radius, is uniformly distributed.

The location privacy must be measured and quantified with respect to the accuracy of the location measurement, which means that the more accuracy the measurement is, the less privacy it achieves. To balance the accuracy and privacy, the metric relevance is defined to measure the location accuracy and location privacy in the measurement.

Relevance [18] Let $A_i =< x_i, y_i, A_i >$ be the location measurement and r_o be the area radius that would be produced if the optimal accuracy is guaranteed. The relevance associate with A_i is the ratio $R_i = r_o^2/r_i^2$.

It is obvious that R_i models the relative accuracy loss of the measurement with respect to the optimal accuracy r_o, such that the location privacy of A_i can be denoted as 1-R_i. The final relevance R_f is defined as the location measurement that should not exceed. The relevance should be specified by the users based on their individual preferences using the concept of minimum distance. For example, a user can define the privacy preference as "10 meters", indicating that the user can be located with the accuracy not larger than 10 meters. This privacy preference also results in the accuracy degradation, which needs to be defined for privacy reason.

Accuracy Degradation [18] Let A_i be the location measurement with initial relevance R_i and final relevance R_f defined by the user. The accuracy degradation to A_i is the ratio $\lambda = R_f/R_i$.

Given a location measurement and an accuracy degradation, the location measurement can be obfuscated in a way that the output satisfies the privacy preference R_f defined by the user.

Obfuscation [18] Let (x_u, y_u) be the user's real position, A_i with relevance R_i be a location measurement, and R_f be the final relevance to be satisfied. A_i is transformed into an obfuscated area A_f such that the following conditions are satisfied:

- A_f have relevance R_f;
- $P((x_u, y_u) \in A_f) > 0$.

Basic obfuscation operation computes an obfuscated area A_f with relevance R_f based on the location measurement A_i with relevance R_i. Therefore, the obfuscation operation can be defined as follows:

Obfuscation Operation [18] Let $\mathscr{A}$ be the set of circular areas, an obfuscation operation $\mathscr{O} : \mathscr{A} \times (0, 1] \times (0, 1] \rightarrow \mathscr{A}$ takes the circular area A_i and two relevances R_i and R_f as input, where R_i is the relevance associated with A_i and $R_f < R_i$ is the final relevance to be satisfied, and outputs as an obfuscated area A_f such that:

- A_f have relevance R_f;
- $A_f \cap A_i \neq \varnothing$.

2.2.2 *Differential Location Privacy*

Location Based Service (LBS) [10] is a popular mobile service for the development and adoption of smartphones. The smartphones embedded with GPS modules have powerful computation and communication capabilities. They can sense users' locations and upload them to an LBS provider which will further process these location information. For example, LBS enables users to share their real-time locations [19], explore places (e.g., restaurants, cinemas) [20], and publish their trajectories [21].

Although LBSs render users a great amount of benefits, still the users' locations are exposed to the server provider and it raises critical privacy concerns for users.

Firstly, the location information is considered as private information of users since they do not need strangers to know where they are. Secondly, after the location information is leaked, it can be readily linked to users' sensitive activities that uses want to protect, such as regular visits to a gym and a sudden visit to a clinic. Through a continuous collecting and processing of these data, an LBS provider can infer a user's home address, diet preference, insurance condition, etc.

We now introduce two applications in LBSs as two utilizations of differential privacy.

The first one is securing LBS applications with geo-indistinguishability guarantees without sacrificing the quality of the application results [22]. It first proposes a formal definition for LBSs. Then it presented a randomized technique to upload users' location information to provide them location services while satisfying the proposed privacy definition.

Specifically, it considers a Paris user who queries an LBS provider for nearby restaurants in a privacy-preserving way. The proposed concept is geo-indistinguishability and the main idea is that for any radius $R > 0$, the user has ϵR privacy within R, which means that the privacy level L is proportional to the chosen radius. A privacy requirement is a constraint on distributions $F(x)$, $F(y)$ conducted on two points $x, y.d(\cdot, \cdot)$ is the Euclidean distance. Satisfying L-privacy within radius R means that for any x, y with $d(x, y) < R$, $d_{\mathcal{P}}(F(x), F(y))$ is at most L. The formal definition is as follows:

[Geo-Indistinguishability] A mechanism F satisfies ϵ-geo-indistinguishability if and only if for all x, y:

$$d_{\mathcal{P}}(F(x), F(y)) \leq \epsilon d(x, y). \tag{2.6}$$

Then, it defines a mechanism to achieve geo-indistinguishability for the continuous plane. When a user has a real location x, the mechanism randomly generates a Laplace noise and adds to x to obtain a new point x'. By doing so, the probabilities of reporting a point in an area around x', when the real locations are x and y, differ at most by $e^{-\epsilon d(x,y)}$. Next, it discretizes the mechanism by mapping each location point to the closest location point in the discrete domain. Finally, it truncates the discretized mechanism which generates location points only within a specific area to fully satisfy the defined geo-indistinguishability. In addition, when a user is performing multiple queries to the LBS server, one way of guaranteeing geo-indistinguishability is to independently generate new locations for the user' each location. As a result, N queries via a mechanism providing ϵ-geo-indistinguishability achieve $N\epsilon$-geo-indistinguishability.

The second one is achieving differential privacy in trajectory data publication[23]. It first points out that existing work related to trajectory data publication cannot full achieve differential privacy because the original noises randomly drawn from Laplace distribution are unbounded. These noises will lead to meaningless trajectory release and the trajectory count on each road must be considered at the same time. Then it proposes a new trajectory data publication algorithm based on a bounded Laplace noise generation algorithm.

Specifically, it utilizes k-means clustering to partition the original location space into groups, and obtains a partition result $\widetilde{pa}$ and a candidate partition set τ. A utility function $U : DB \times tau \rightarrow R$ is defined to assign a utility score to each partition pa and the formal definition is:

$$U(D, p) = \frac{MeanDist(\widetilde{pa})}{MeanDist(pa)} \tag{2.7}$$

where $MeanDist(pa)$ is the mean distance of any two trajectories in partition pa. Then it leverage the exponential mechanism[24] to select one partition pa_i following the probability: $\frac{exp(\frac{\epsilon}{2\Delta U}U(DB,pa_i))}{\sum_{pa\in\tau} exp(\frac{\epsilon}{2\Delta U}U(DB,pa))}$. Next, it includes the generalized trajectories with some randomly chosen original trajectories to form a generalized trajectory database, and generates bounded Laplace noises to be added to the counts of generalized trajectories. Finally, it enforces a consistency constraint on the noisy counts to form a more accurate and consistent trajectory data publication by using constrained inference technique.

2.2.3 Privacy-Preserving Location Matching

In IoT applications, such as mobile crowdsensing and location-based services, a third party may be willing to test whether two locations provided by users are matched (equal) or not for offering location-related services to users [25]. For example, the sensing tasks released by the customers define the sensing areas where the mobile users collect data, and the mobile users are required to upload their location information, such that the service provider is able to allocate the sensing tasks based on the sensing areas of tasks and locations of mobile users, and thereby the data quality reported by mobile users can be improved. In doing so, the geocast areas of sensing tasks are represented as a vector $L = \{l_1, l_2, \cdots, l_n\}$, and the location of a worker $\mathbb{U}$ is supported to be l. The service provider learns whether $l \in L$ with no knowledge about l and L. Thus, the service provider recommends the task with L to $\mathbb{U}$, if $l \in L$ [26].

Service Setup The service provider setups the whole service by defining the system parameters. It selects a security parameter k and determines the prime order q of bilinear groups. Generally, $k = 160$ or 256. Let G be a cyclic group, and G_T be a multiplicative group with the order q. P is the generator of G. $\hat{e}$ is a bilinear pairing $\hat{e} : G \times G \rightarrow G_T$. $C = E_{AES}(K, M)$ and $M = D_{AES}(K, C)$ are the encryption and decryption algorithms of AES. The service provider randomly chooses $Q \in G$ and two cryptographic hash functions $H_1 : \{0, 1\}^* \rightarrow Z_q^*$ and $H_2 : \{0, 1\}^* \rightarrow G$. The public parameter is $gp = \{q, G, G_T, \hat{e}, P, Q, H_1, H_2\}$. In addition, the service provider defines the service geographic region for customers, i.e., the points of interest in the regions, such as shopping malls, museums, plazas and buildings

A worker $\mathbb{U}$ generates a public-private key pair by randomly picking $s \in Z_q^*$ as the privacy key to compute $P_{pub} = sP$ as the public key. A customer $\mathbb{C}$ randomly picks $v \in Z_q^*$ as the secret key and calculates the public key as $V_{pub} = vP$.

Task Releasing When $\mathbb{C}$ is willing to collect data, it generates a sensing task $ST = (Cont, Expt, L)$, which indicate the content (what to sense), the expiration time (when to sense) and the geocast area (where to sense). $\mathbb{C}$ randomly chooses num as the identifier. The geocast area L is a set of points of interest from which $\mathbb{C}$ needs to collect and analyze data, denoted as $L = \{l_1, l_2, \cdots, l_n\}$. To prevent $L = \{l_1, l_2, \cdots, l_n\}$, $\mathbb{C}$ computes a series of encrypted points of interest as follows:

1. Randomly pick $k \in Z_q^*$ and compute $K_{pub} = kP$.
2. Randomly choose $\gamma \in Z_q^*$ to calculate $C_1 = \gamma P$ and $h = H_1(num, cert_c, \hat{e}(K_{pub}, Q)^{\gamma})$.
3. For $i = \{1, \cdots, n\}$, compute $x_i = H_1(l_i)$, and

$$f_i(x) = \prod_{1 \le j \ne i \le n} \frac{x - x_j}{x_i - x_j}$$
$$= a_{i,1} + a_{i,2}x + \cdots + a_{i,n}x^{n-1}.$$

   where $a_{i,1}, \cdots, a_{i,n} \in Z_p^*$.
4. For $i = \{1, \cdots, n\}$, randomly choose $\alpha_i \in Z_q^*$, and compute $y_i = \alpha_i^{-1}\gamma$, and $U_i = \sum_{j=1}^{n} a_{j,i}\alpha_j K_{pub}$.
5. For $i = \{1, \cdots, n\}$, compute $X_i = H_2(l_i \mid\mid num)$ and $R_i = \sum_{j=1}^{n} a_{j,i} y_j X_j$.
6. Set $L = \{l_1, l_2, \cdots, l_n\}$ of $\mathbb{C}$ as $C = (R_1, \cdots, R_n, U_1, \cdots, U_n, C_1, h)$.

$\mathbb{C}$ sends $(num, Cont, Expt, C)$ to the service provider.

Task Recommendation When $\mathbb{U}$ participates in crowdsensing activities, $\mathbb{U}$ interacts the service provider to retrieve the recommended spatial task:

1. $\mathbb{U}$ calculate a location trapdoor $T_l = (T_1, T_2)$ as $T_1 = H_1(l)$, and $T_2 = k(Q + H_2(l \mid\mid num))$ based on the location l. $\mathbb{U}$ forwards $(cert_u, T_l)$ to the service provider.
2. The service provider utilizes T_l to calculate

$$\lambda = R_1 + R_2T_1 + \cdots + R_nT_1^{n-1} \pmod q,$$

$$\nu = U_1 + U_2T_1 + \cdots + U_nT_1^{n-1} \pmod q$$

   and then verifies whether

$$h \stackrel{?}{=} H_1(num, cert_c, \frac{\hat{e}(C_1, T_2)}{\hat{e}(\nu, \lambda)}). \tag{2.8}$$

   If the equation holds, $\mathbb{U}$'s location l is in L, the service provider sends the task to $\mathbb{U}$.

2.3 Data Privacy-Enhancing Techniques

2.3.1 Randomization

Given an input m_i, each user i can generate a blinding factor r_i and publishes a blinded value $m_i' = m_i + r_i \mod M$, where M is the integer modulus that efficiently represents the message space. Thus, the input m_i is randomized and protected by the random value r_i. This approach is widely used in computing data statistics. If $\sum_i^n r_i = 0$, where n is the number of users, $\sum_i^n m_i' = \sum_i^n m_i = 0$.

2.3.2 Data Encryption

Data encryption is a widely-used method to keep data confidential in communication systems. The basic structure of a public-key encryption scheme (PKE) consists of three algorithms [27]:

- $PKE.KeyGen(\lambda)$ This algorithm takes the security parameter λ as input and outputs a public key/secret key pair (PK, SK). The structure of PK and SK depend on the particular scheme.
- $PKE.Encrypt(PK, format, L, M)$ This algorithm takes the public key PK, a format, a label L, and a cleartext M as input, and outputs a ciphertext C. The format is optional, and its structure and meaning depends on the particular encryption scheme.
- $PKE.Decrypt(SK, L, C)$ This algorithm takes a secret key SK, a label L, and a ciphertext C as input, and outputs a cleartext M.

Lifted ElGamal [28]. The Lifted ElGamal scheme can be proved to be secure under Decisional Diffie-Hellman assumption and is composed by three algorithms: key generation, encryption and decryption.

- *Key Generation* Given a secure parameter κ, pick a big prime p. Let $\mathbb{G}, \mathbb{G}_1$ be groups of order p and $\hat{e} : \mathbb{G} \times \mathbb{G} \to \mathbb{G}_1$ be a bilinear map. Choose a random generator $g \in_R \mathbb{G}$ and a random value $x \in_R \mathbb{Z}_p^*$ to compute $h = g^x \in \mathbb{G}$. The public key is $\mathscr{PK} = (p, \mathbb{G}, \mathbb{G}_1, e, g, h)$ and the private key is $\mathscr{SK} = x$.
- *Encryption* Assume the plaintext space consists of elements in the set $\{0, 1, \cdots, T\}$ with $T \leq p$. To encrypt a plaintext message m using the public key $\mathscr{PK}$, pick a random $r \in \mathbb{Z}_p^*$ and calculate $C_1 = g^r \in \mathbb{G}$, $C_2 = g^m h^r \in \mathbb{G}$ as the ciphertext.
- *Decryption* To decrypt a ciphertext (C_1, C_2) using the private key $\mathscr{SK} = x$, compute $C^* = C_2 C_1^{-x}$. To cover m, it suffices to compute the discrete log of C^* base g using Pollard's lambda method [21].

The cryptosystem is additively homomorphic. Specifically, give two ciphertexts (C_1, C_2) and (C_1', C_2') of plaintexts m and m' respectively, anyone can compute the ciphertexts of $m'' = m+m'$ by computing the product $C_1'' = C_1 C_1' g^r, C_2'' = C_2 C_2' h^r$ for a random $r \in \mathbb{Z}_p^*$.

Proxy Re-Encryption [29]. Proxy Re-encryption is a special public key encryption with a desirable property that a semi-trusted proxy enables to convert a ciphertext for Alice into a ciphertext for Bob without seeing the underlying plaintext, given a proxy re-encryption key. Thanks to this promising property, it has been widely employed in data sharing scenarios. The proxy re-encryption scheme is proposed by Ateniese et al. [29], the details of which are as follows:

- KeyGen($\cdot$) Alice picks a random value $a \in \mathbb{Z}_p$ as the secret key sk_a and compute the public key $pk_a = g^a$.
- RKeyGen(sk_a, pk_b) Alice delegates to Bob by sending the re-encryption key $rk_{A \to B} = g^{b/a}$ to a proxy by using Bob's public key.
- Encrypt(m, pk_a) To encrypt a message $m \in \mathbb{G}_T$ under pk_a, Alice chooses a random value $k \in \mathbb{Z}_p$ to compute $c_a = (g^{ak}, m\hat{e}(g, g)^k)$.
- Re-Enc($c_a, rk_{A \to B}$) The proxy can change the ciphertext c_a into a ciphertext c_b for Bob with $rk_{A \to B}$. From c_a, the proxy calculates $\hat{e}(g^{ak}, g^{b/a}) = \hat{e}(g, g)^{bk}$ and releases $c_b = (\hat{e}(g, g)^{bk}, m\hat{e}(g, g)^k)$.
- Decrypt (c_b, sk_b) Bob enables to decrypt c_b to obtain m as $m = m\hat{e}(g, g)^k/(\hat{e}(g, g)^{bk})^{1/b}$.

2.3.3 Differential Privacy

Differential privacy is a definition of privacy that is used to measure privacy in a quantitative way. ϵ-differential privacy A function F satisfies ϵ-differential privacy [10, 19] if for all databases DB_1 and DB_2 differing on at most one item, and all outputs $O \in Range(F)$,

$$Pr[F(DB_1) \in O]e^{\epsilon} \times Pr[F(DB_1) \in O], \tag{2.9}$$

where the possibility is taken over the randomness of F.

The first concept is privacy budget ϵ and it indicates the privacy protection strength which increases when ϵ decreases [20, 21].

The second concept is global sensitivity [22] and it captures the maximum difference when queried on two neighboring databases. Global Sensitivity: For any function $F : DB \to R^d$, the sensitivity ΔF of function F is:

$$\Delta F = \max_{d(DB_1, DB_2)=1} |F(DB_1) - F(DB_2)|. \tag{2.10}$$

Two common methods to realize differential privacy are Laplace mechanism [19] and exponential mechanism [23].

Laplace mechanism is used for functions whose query outputs are real numbers. Its inputs are database D, function F and privacy budget ϵ. Laplace mechanism draws a random Laplace noise from a Laplace distribution and adds this noise to the query output. Specifically, the Laplace distribution has a probability density function $Pr[x|\mu, a] = \frac{1}{2a}e^{(-|x-\mu|)/a}$ with location parameter μ and scale parameter a.

Exponential mechanism is used for functions whose query outputs are not real. It first gives a utility function U to match a utility score to every query output and then assigns higher probabilities to be selected to these query outputs with higher utility scores in an exponential way.

There are two common properties of differential privacy when it is utilized in sequential composition and parallel composition [24].

First, when several functions which all give ϵ-differential privacy work in a sequence on a same database, this combination will give $\sum_i \epsilon_i$-differential privacy. Second, if the above functions are conducted on different databases, then the overall privacy budget is determined by the biggest privacy budget.

2.3.4 *Zero-Knowledge Proof*

Zero-knowledge proof [30] is generating excitement recently due to its great potential to enhance security and privacy for online services, especially blockchain-based business process management and crypto concurrency (e.g. Corda and Bitcoin). In this chapter, we give a brief introduction to the fundamentals of zero-knowledge proof technique and present some promising use cases.

Zero-knowledge proof allows a prover $\mathscr{P}$ with a secret w to convince a prover $\mathscr{V}$ a relation R is true regarding to some common reference string s while keeping the secret w private to the verifier. Formally, a verifier convinces the prover that he/she knows some w and $(w, x) \in R$. In other words, zero-knowledge proof enable a verifier to validate the truth of something without leaking contents of the truth to the verifier. A zero-knowledge proof system must satisfy the following three properties:

- **Completeness** An honest verifier $\mathscr{V}$ (which means $\mathscr{V}$ correctly follows the protocol) will be convinced by an honest prover $\mathscr{P}$ if the statement is true.
- **Soundness** If a statement is false, no dishonest prover $\mathscr{P}$ can convince the verifier $\mathscr{V}$ the statement is true, except with some negligible probability.
- **Zero Knowledge** A verifier learns nothing other than the correctness of the statement. In other words, it is sufficient for the verifier who only knows the statement to imagine a scenario that shows the prover knows the secret. In doing so, the verifier must be able to construct a simulator, that takes into only the statement to be proved and generates a transcript that simulates an interaction between the honest verifier and prover.

Zero-knowledge proof systems can be instantiated for different settings of relations. Generally speaking, zero-knowledge proof exists for all of NP-complete language, e.g. Circuit Satisfiability. For efficiency concerns, we can instantiate zero-knowledge proof system for a specific language. An important observation is that many public-key cryptography protocols are based on finite abelian groups. Let $\mathbb{G}_1$, $\mathbb{G}_2$, and $\mathbb{G}_T$ be three cyclic groups with prime order p and an efficiently computable map: $e : \mathbb{G}_1 \times \mathbb{G}_2 \rightarrow \mathbb{G}_T$. We first look at languages in discrete logarithm (DL) setting, where knowledge that is to be proven is integers from $\mathbb{Z}_p$.

Zero-knowledge for Discrete Logarithm We usually use the notation introduced by Camenisch, Kiayias, and Yung. For example, we denote

$$\mathbf{PoK}\{(a, b, c) : Y = g_1^a H^b \wedge \hat{Y} = \hat{g}_1^a \hat{H}^c\} \tag{2.11}$$

which is a zero-knowledge proof of knowledge of integers a, b, c such that $Y = g_1^a H^b$ and $\hat{Y} = \hat{g}_1^a \hat{H}^c$ holds, where a, b, c are integers from $\mathbb{Z}_p$ that are to be proven and $Y, g_1, \hat{Y}, \hat{g}_1$ are group elements from $\mathbb{G}_1$ that are publicly known to the verifier. A general construction of zero-knowledge proof of knowledge is denoted as

$$\mathbf{PoK}\{(\alpha_1, \alpha_2 \ldots \alpha_m) | \bigwedge_{i=1}^{n} X_i = f_i(\alpha_1, \alpha_2 \ldots \alpha_m)\} \tag{2.12}$$

where $(\alpha_1, \alpha_2 \ldots \alpha_m) \in \mathbb{Z}_p^m$ is the knowledge that satisfies some publicly computable statements f_i for some public values X_i. In DL setting, we can initiate the proof system by a non-interactive preimage proof combining the most common $\sum$ protocol and Fiat-Shamir heuristic [31].

Zero-knowledge Proof for Bilinear Groups In DL setting, we mainly focus on knowledge that is integers from $\mathbb{Z}_p$. For more general constructions, we need to consider that knowledge is group elements from $\mathbb{G}_1$ or $\mathbb{G}_2$. In [32], Groth and Sahai proposed a general instantiation for practical cryptographic protocols based on bilinear groups, which is well known as GS proof. The generality of their work can be extended to two directions. First, they formulated their constructions over commutative rings with an efficient bilinear mapping that captures all existing cryptographic bilinear groups. Second, all operations in a context of a bilinear group are taken into consideration in their proposed framework, such as additions in $\mathbb{G}_1$ and $\mathbb{G}_2$, addition or multiplication of scalars, and any operations regarding a bilinear paring. Third, the knowledge to be proven in GS setting can include group elements in $\mathbb{G}_1$ and $\mathbb{G}_2$ and integers in $\mathbb{Z}_p$.

Zero-knowledge proof system for discrete logarithm and bilinear groups can be used to construct many interesting applications with specific statements. In this paper, we do not distinguish zero-knowledge proof and zero-knowledge argument. The two terms are used interchangeably. Here, we present some interesting examples.

- Zero-knowledge proof for general arithmetics Zero-knowledge proofs in DL or GS setting can be used to construct witness that satisfies a specific arithmetic circuit. In [33], Bootle et al. proposed an efficient and honest verifier zero-knowledge argument system for general arithmetic circuit satisfiability in discrete logarithm setting with a logarithmic communication complexity.
- Zero-knowledge proof for polynomial evaluations In [34], Bootle and Groth proposed a general framework for evaluating polynomials with low degree. Specifically, they transfer the evaluation of polynomials to an arithmetic circuit and a witness ia assignments that satisfy the circuits. Moreover, Bootle and Groth

designed a batch verification method to improve efficiency in verification and reduce the proof size.

- Zero-knowledge proof for membership Starting from zero knowledge argument for arithmetic circuit satisfiability, there are some more specific applications. For example, a prover can construct a zero-knowledge witness showing that the prover knows some secret that is a member of a large data set. This is well known as a membership zero-knowledge argument or 1-out-of-N membership proof formally. Further, 1-out-of-N membership argument can be extended to achieve zero-knowledge proof of a Hamming weight [35]. Specifically, a prover can prove to a verifier that the Hamming weight of a vector is less than a publicly known threshold without leaking the vector to the verifier.

SNARGs When generating zero-knowledge arguments for complex statements such as polynomial evaluations in discrete logarithm or GS setting, a practical issue is that the size of the proof is usually increasing with the size of the statements (e.g. degree of the polynomials). This makes the zero-knowledge argument infeasible for real-world applications since the proof size can be even a few gigabytes. On addressing this issue, research efforts are being directed to succinct non-interactive arguments (SNARGs) for verifying NP-complete language with much less complexity in computation and communication. In [36], Boneh et al. proposed quasi-optimal SNARGs for Boolean circuit satisfiability where the proof complexity and proof size is $\hat{O}(|C|) + poly(\lambda, log|C|)$, where λ is the security parameter and $|C|$ is the size of the Boolean circuit. Extending SNARGs for Boolean circuit, Ben-Sasson et al. [37] proposed a SNARKs for verifying a C program succinctly in zero knowledge. In specific, the authors constructed TinyRam, a nondeterministic random-access machine for C programs.

Post-Quantum Zero Knowledge Almost a decade ago, we knew that a powerful quantum computer can break nearly all existing public key systems. Thus, zero-knowledge proof system in traditional bilinear groups is prone to attacks in post quantum era, which motivates invention of cryptographic primitives with post quantum security. In [38], Chase et al. proposed a post-quantum zero-knowledge proof system based on symmetric primitives along with the "MPC-in-the-head" paradigm. Specifically, they constructed a proof system for general circuit verification and instantiated the proof for a common hash function, where a prover proves to a verifier that he knows a secret x that is the preimage of y such that $y = SHA256(x)$. However, communication overheads and proof size increase significantly for post-quantum zero-knowledge proof system.

Zero-knowledge for Blockchain and Smart Contract A blockchain is a global leger maintained by a P2P network with some-predefined consensus protocol such as Proof-of-Work (PoW) or Proof-of-Stake (PoS). A blockchain contains a continuously growing list of records, called blocks, which are linked and secured using cryptographic hash functions and signatures. Ideally, a blockchain can record transactions between two anonymous parties in an immutable and transparent way. That is, anyone can verify the current state of the blockchain and the transaction history on the public ledger. If we regard blockchain as immutable and transparent decentralized conceptual party that can be trusted for availability and trustness, we can build smart contracts atop of the blockchain [39]. Specifically, a smart contract allows mutually distrustful parties to agree on and execute a pre-defined contract. Applications built upon blockchain and smart contracts have been envisioned with great market value, such as Bitcoin and Ethereum. Since all information on blockchain and smart contract is public, privacy of transactions and contracts are not preserved: the sender, the receiver, and the transaction amount. To address the confidentiality and privacy concerns of the transactions, existing literature adopts zero-knowledge proof to hide the specific input to the blockchain and smart contract. For example, a party can send inputs to a smart contract and demonstrate the inputs fulfill the contract requirements in a zero-knowledge manner.

2.4 Summary

We have reviewed the state-of-the-art privacy-enhancing techniques which are utilized to achieve identity, location and data privacy preservation.

References

1. M. Raya and J.-P. Hubaux, "Securing vehicular ad hoc networks," *Journal of Computer Security - Special Issue on Security of Ad-hoc and Sensor Networks*, vol. 15, no. 1, pp. 39–68, 2007.
2. P. Samarati and L. Sweeney, "Protecting privacy when disclosing information: k-anonymity and its enforcement through generalization and suppression," SRI International, Tech. Rep., 1998.
3. L. Sweeney, "k-anonymity: A model for protecting privacy," *International Journal of Uncertainty, Fuzziness and Knowledge-Based Systems*, vol. 10, no. 5, pp. 557–570, 2002.
4. ——, "Achieving k-anonymity privacy protection using generalization and suppression," *International Journal of Uncertainty, Fuzziness and Knowledge-Based Systems*, vol. 10, no. 5, pp. 571–588, 2002.
5. A. Machanavajjhala, J. Gehrke, D. Kifer, and M. Venkitasubramaniam, "l-diversity: Privacy beyond k-anonymity," in *Proceedings of the 22nd International Conference on Data Engineering*, 2006, p. 24.
6. N. Li, T. Li, and S. Venkatasubramanian, "t-closeness: Privacy beyond k-anonymity and l-diversity," in *Proceedings of the 23rd International Conference on Data Engineering*, 2007, pp. 106–115.
7. D. Chaum, "Untraceable electronic mail, return addresses, and digital pseudonyms," *Communications of the ACM*, vol. 24, no. 2, pp. 84–88, 1981.
8. C. Park, K. Itoh, and K. Kurosawa, "Efficient anonymous channel and all/nothing election scheme," in *Workshop on the Theory and Application of Cryptographic Techniques*, vol. 765, 1993, pp. 248–259.
9. P. Golle, M. Jakobsson, A. Juels, and P. F. Syverson, "Universal re-encryption for mixnets," in *Proceedings of the Cryptographer's Track at the RSA Conference*, vol. 2964, 2004, pp. 163–178.

10. C. A. Neff, "A verifiable secret shuffle and its application to e-voting," in *Proceedings of the 8th ACM Conference on Computer and Communications Security*, 2001, pp. 116–125.
11. O. Pereira and R. L. Rivest, "Marked mix-nets," in *Financial Cryptography and Data Security*, 2017, pp. 353–369.
12. F. Zhang and K. Kim, "Id-based blind signature and ring signature from pairings," in *Proc. of AsiaCrypto*, 2002, pp. 533–547.
13. D. Boneh and M. Franklin, "Identity-based encryption from the weil pairing," in *Annual international cryptology conference*, 2001, pp. 213–229.
14. J. Camenisch and M. Stadler, "Efficient group signature schemes for large groups," in *Proc. of CRYPTO*, 1997, pp. 410–424.
15. D. Pointcheval and O. Sanders, "Short randomizable signatures," in *Proc. of CT-RSA*, 2016, pp. 111–126.
16. X. Lin, X. Sun, P.-H. Ho, and X. Shen, "Gsis: A secure and privacy-preserving protocol for vehicular communications," *IEEE Transactions on vehicular technology*, vol. 56, no. 6, pp. 3442–3456, 2007.
17. H. Xiong, Z. Qin, and F. Li, "An anonymous sealed-bid electronic auction based on ring signature." *IJ Network Security*, vol. 8, no. 3, pp. 235–242, 2009.
18. C. A. Ardagna, M. Cremonini, S. D. C. di Vimercati, and P. Samarati, "An obfuscation-based approach for protecting location privacy," *IEEE Transactions on Dependable and Secure Computing*, vol. 8, no. 1, pp. 13–27, 2011.
19. C. Dwork, F. McSherry, K. Nissim, and A. Smith, "Calibrating noise to sensitivity in private data analysis," in *Proc. of TCC*, 2006, pp. 265–284.
20. C. Dwork, "A firm foundation for private data analysis," *Communications of the ACM*, vol. 54, no. 1, pp. 86–95, 2011.
21. C. Dwork, A. Roth *et al.*, "The algorithmic foundations of differential privacy," *Foundations and Trends® in Theoretical Computer Science*, vol. 9, no. 3–4, pp. 211–407, 2014.
22. A. Blum, K. Ligett, and A. Roth, "A learning theory approach to noninteractive database privacy," *Journal of the ACM (JACM)*, vol. 60, no. 2, p. 12, 2013.
23. F. McSherry and K. Talwar, "Mechanism design via differential privacy," in *Proc. of FOCS*, 2007, pp. 94–103.
24. F. D. McSherry, "Privacy integrated queries: an extensible platform for privacy-preserving data analysis," in *Proceedings of the 2009 ACM SIGMOD International Conference on Management of data*, 2009, pp. 19–30.
25. J. Ni, A. Zhang, X. Lin, and X. Shen, "Security, privacy, and fairness in fog-based vehicular crowdsensing," *IEEE Communications Magazine*, vol. 55, no. 6, pp. 146–152, 2017.
26. A. Alamer, J. Ni, X. Lin, and X. Shen, "Location privacy-aware task recommendation for spatial crowdsourcing," in *Proc. of WCSP*, 2017, pp. 1–6.
27. M. Bellare, A. Desai, D. Pointcheval, and P. Rogaway, "Relations among notions of security for public-key encryption schemes," in *Proc. of Crypto*, 1998, pp. 26–45.
28. J. Liu, N. Asokan, and B. Pinkas, "Secure deduplication of encrypted data without additional independent servers," in *Proc. of ACM CCS*, 2015, pp. 874–885.
29. G. Ateniese, K. Fu, M. Green, and S. Hohenberger, "Improved proxy re-encryption schemes with applications to secure distributed storage," *ACM Transactions on Information and System Security (TISSEC)*, vol. 9, no. 1, pp. 1–30, 2006.
30. S. Goldwasser, S. Micali, and C. Rackoff, "The knowledge complexity of interactive proof systems," *SIAM Journal on computing*, vol. 18, no. 1, pp. 186–208, 1989.
31. A. Fiat and A. Shamir, "How to prove yourself: Practical solutions to identification and signature problems," in *Proc. of CRYPTO*, 1986, pp. 186–194.
32. J. Groth and A. Sahai, "Efficient non-interactive proof systems for bilinear groups," in *Annual International Conference on the Theory and Applications of Cryptographic Techniques*. Springer, 2008, pp. 415–432.
33. J. Bootle, A. Cerulli, P. Chaidos, J. Groth, and C. Petit, "Efficient zero-knowledge arguments for arithmetic circuits in the discrete log setting," in *Annual International Conference on the Theory and Applications of Cryptographic Techniques*. Springer, 2016, pp. 327–357.

34. J. Bootle and J. Groth, "Efficient batch zero-knowledge arguments for low degree polynomials," in *IACR International Workshop on Public Key Cryptography*. Springer, 2018, pp. 561–588.
35. I. Damgård, J. Luo, S. Oechsner, P. Scholl, and M. Simkin, "Compact zero-knowledge proofs of small hamming weight," in *IACR International Workshop on Public Key Cryptography*. Springer, 2018, pp. 530–560.
36. D. Boneh, Y. Ishai, A. Sahai, and D. J. Wu, "Quasi-optimal snargs via linear multi-prover interactive proofs," in *Annual International Conference on the Theory and Applications of Cryptographic Techniques*. Springer, 2018, pp. 222–255.
37. E. Ben-Sasson, A. Chiesa, D. Genkin, E. Tromer, and M. Virza, "Snarks for c: Verifying program executions succinctly and in zero knowledge," in *Advances in Cryptology–CRYPTO 2013*. Springer, 2013, pp. 90–108.
38. M. Chase, D. Derler, S. Goldfeder, C. Orlandi, S. Ramacher, C. Rechberger, D. Slamanig, and G. Zaverucha, "Post-quantum zero-knowledge and signatures from symmetric-key primitives," in *Proceedings of the 2017 ACM SIGSAC Conference on Computer and Communications Security*. ACM, 2017, pp. 1825–1842.
39. A. Kosba, A. Miller, E. Shi, Z. Wen, and C. Papamanthou, "Hawk: The blockchain model of cryptography and privacy-preserving smart contracts," in *Security and Privacy (SP), 2016 IEEE Symposium on*, 2016, pp. 839–858.

Chapter 3
Identity Privacy Protection in Smart Parking Navigation

Due to the large volumes of modern vehicles in metropolises, finding a vacant parking space has become an irritating and frustrating problem for drivers, particularly in a congested area, such as sport centers, shopping malls, and downtown [1]. The extra traffic cruising for parking spaces brings serious social problems, including traffic congestion, vehicle accident, fuel waste, and air pollution [2]. Although Google Maps and portable navigators assist drivers to discover parking garages in destinations, drivers are then faced with a problem that no vacant parking space are available [3].

Vehicular ad hoc network (VANET) enables each vehicle equipped with an onboard unit (OBU) to interact with the nearby vehicles and with the roadside units (RSUs) [4, 5]. It can provide various advanced applications for road safety improvement and driving experience enrichment, where smart parking navigation provides real-time parking navigation service to guide vehicles to available parking spaces [6]. In smart parking navigation, OBU sends a parking query to the nearby RSUs (i.e., fogs) for parking space discovery around its destination and drive to the vacant parking spot based on the up-to-date parking information. VANET-based navigation has the advantage that drivers conveniently utilize OBUs to enjoy real-time parking navigation services and arrive vacant parking spaces within short delay and low fuel cost.

Security and privacy are preliminary concerns for drivers in VANETs [7], since the infrastructure may be confronted with various cyber attacks, including impersonation attacks, forgery attacks and global eavesdropping attacks [8, 9]. The leakage of identity and location information is a huge concern for drivers, which triggers numerous controversies on track exposure [10]. Current navigation applications, such as Google Maps, Apple Maps and Baidu Maps, collect drivers' locations and destinations [11], resulting in the leakage of drivers' trajectory and the exposure of their personal habits. OBUs frequently communicate with RSUs (fogs) to deliver parking queries, including current locations and destinations, to acquire navigation information to parking spaces. Thus, curious entities, including fogs and

X. Lin et al., *Privacy-Enhancing Fog Computing and Its Applications*,
SpringerBriefs in Electrical and Computer Engineering,
https://doi.org/10.1007/978-3-030-02113-9_3

clouds, learn the driving patterns of vehicles and predict the drivers' locations at a future time, or even identify sensitive information about drivers, e.g., references, home addresses, social relationships, health conditions, and political affiliations, based on the visiting frequency of specific spots. in addition, the exposure of vehicles' locations might give convenience to car thieves, who trace the vehicles several days before taking action and prefer to steal cars in quiet places [12]. Therefore, location privacy preservation is critical for the wide acceptance of smart parking navigation services to the public.

One common method to preserve location privacy is to keep the anonymity of drivers [13, 14]. Once the drivers are kept anonymous, attackers cannot identify the drivers or link several location information to reconstruct a specific driver's trajectory. Unfortunately, if the identities are hidden, returning navigation results to the correct vehicles becomes challenging. To address this challenge, Chim et al. [15] assume that the vehicle keeps the connection with the fog after delivering the navigation query until receiving the result. This approach is quite challenging to perform in reality, particularly, when the vehicle moves at high speeds. The handover of V2R connections and signals blocking of buildings increase the disconnection probability of a querying vehicle [16]. As a result, the delivery probability of navigation results is limited. In addition, full anonymity is not perfect as it is difficult to charge drivers for service access or unlicensed or unqualified drivers' identification, who obtain too many demerit points on driving records. Thereby, the drivers' identities should be recovered for service charging and unqualified drivers identification [17].

In this chapter, we propose a Privacy-preserving Smart PAking Navigation scheme (P-SPAN) to offer secure smart parking navigation services for drivers [18]. We observe that most of drivers utilize GPS navigation systems, thus the route from the source to a destination is determined. Thereby, the driving-through RSUs (i.e., fogs) for a specific driver can be predicted. In this way, drivers can query vacant parking spaces through vehicle communications and acquire the navigation results from the fog on the way to the destinations, Specifically, the contributions are two-fold:

- To preserve drivers' privacy, P-SPAN achieves conditional identity privacy preservation based on anonymous credentials. To be specific, a registered vehicle sends a parking query to and receives the navigation result without disclosing its real identity, only demonstrating that the vehicle is capable of accessing the navigation service via the anonymous credential. Meanwhile, a trusted authority (TA) can recover the driver's identity for service charging or identifying unqualified drivers.
- An efficient data retrieval mechanism is developed to improve the retrieving probability of navigation results in anonymous vehicular communications based on Bloom filters. The vehicle can retrieve the navigation result from the fogs built on the driving routes from the source to the destination with enhanced success

probability. This method is compatible to the traditional situation, in which the navigation result is returned rapid enough to the vehicle through the queried RSU.

3.1 Problem Statement

We formalize the system model and security threats, and identify security goals.

3.1.1 System Model

A smart parking navigation system has four entities: a cloud, a large number of vehicles, RSUs and a trust authority.

- *Cloud* The cloud, consists of a series of servers and a data center, and offers two kinds of services, parking space management and smart parking navigation. In parking space management service, the servers at parking lots and RSUs beside roads collect and manage the status of parking spots, i.e., vacant, reserved and possessed, charge the parking fee according to the charging policy and outsource the real-time parking information to the data center. The smart parking navigation service is built upon the stored parking data. For instance, the cloud manages the parking lots (red points in Fig. 3.1) around the CN tower and provides the smart parking navigation service to drivers, whose desired destinations are the CN tower.
- *Vehicles* A vehicle is capable of communicating with the nearby vehicles and the fogs using the equipped irreplaceable and temper-proof OBU device. OBUs have computing resources to run simple computations and storage capability for data maintenance, including a certain of read-only memory. Humans with driving licenses are eligible to drive vehicles on roads. Demerit points are added to a driver's licence, if the driver is convicted of breaking certain driving laws. If the driver collects sufficient points, he or she is not qualified to drive in a certain time period.
- *RSUs/Fogs* RSUs, also called fogs, deployed along roadside, interact with each other and with drive-by vehicles. They can interact with the cloud through the Internet. A fog is resource in rich, which means that it has enough computing resource to perform cryptographic operations for the security of information exchange, and storage spaces to store the navigation results for drivers.
- *Trust Authority (TA)* TA is a government agency who administers vehicle registration and driver licensing, e.g., ServiceOntario Centers in Canada and Department of Motor Vehicles (DMV) in USA. It is a fully TA, whose responsibility is to issue public-key certificates to all entities in the system, including the cloud, fogs and vehicles, and recover the drivers' identities in anonymous parking navigation services for service charging or unqualified driver identification.

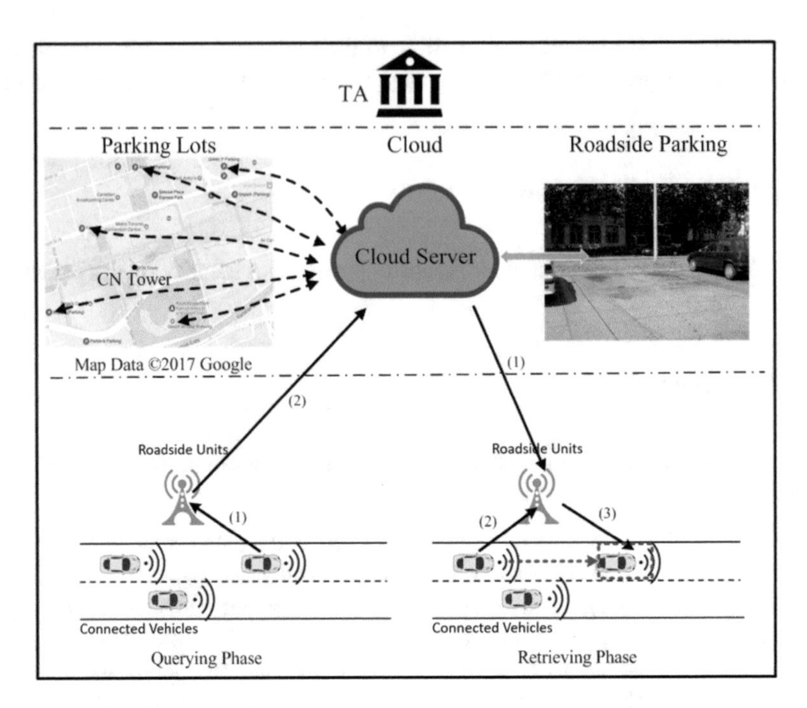

Fig. 3.1 System model of P-SPAN

Fig. 3.1 depicts the system model of smart parking navigation. A driver is required to license at TA and acquires a digital driving licence after passing driving tests, which can be stored on the smart phone or a USB device. The cloud offers the parking space management service to parking lots and roadside parking spaces, and maintains real-time parking information. The cloud also provides the smart parking navigation service to drivers. To access this service, a driver registers the service at the cloud and acquires an anonymous credential for service access. The smart parking navigation consists of two phases: query and retrieval. In the query phase, (1) a vehicle delivers a parking navigation query, including its destination and current location, and delivers it to the nearby fog; (2) The fog receives the parking query and forwards it to the cloud. The cloud discovers an accessible parking space for the driver according to the parking data and the desired destination. In the retrieval phase, (1) the cloud predicts the fogs that the querying vehicle drives through and returns the navigation result to these fogs, and the fogs maintain the navigation result; (2) If the querying vehicle enters the coverage area of a fog, it delivers a retrieving query to the nearby fog; (3) The fog finds the navigation result on storage spaces and returns it to the vehicle, if the fog has the result, otherwise, the vehicle retrieves the result from a future fog. If the recommended parking space is possessed, the cloud updates the navigation result according to the vehicle's location, and the vehicle retrieves the result from the drive-by fogs and obtains the latest navigation result.

3.1.2 Security Threats

The global eavesdroppers capture the transmitting messages exchanged between two entities, such that the eavesdroppers can learn the moving patterns of drivers, predict the locations of drivers at a certain time, and identify preferences and habits of drivers based on the trajectory. Internal attackers are the curious employees in cloud or drivers that would like to have more knowledge about other drivers. Although the cloud would follow the regulations and agreements agreed with drivers, it is interested in drivers' privacy and eager to discover sensitive information from the parking queries or learn a driver's trajectory. These information, having numerous drivers' privacy, may be shared with the cooperators for exploiting hidden values. In addition, even the cloud behaves honestly on data storage, the drivers may believe that their sensitive information would be revealed to the public, due to the frequently accidents of data leakage. Therefore, the cloud is only semi-honest. The vehicles might utilize impersonation attacks to pretend to be legitimate vehicles to access free parking navigation service if this service is charged, or eavesdropping attacks to obtain the navigation result, in case they have the same destination. Besides, they would not share their driving licences or anonymous credentials with other vehicles, since they will be punished once discovered by the TA. In addition, the fogs may be compromised by hackers and they can expose the navigation results stored on storage spaces, or use all sorts of methods to extract private information about drivers via analyzing the forwarding data, e.g., parking queries and navigation results.

3.1.3 Security Goals

To realize privacy-preserving smart parking navigation under the above system model and resist security threats, P-SPAN needs to have the following security goals:

- Identity Authentication A driver is eligible to drive on roads, which means that the driver has a digital driving licence with demerit points, which is less than the threshold.
- Service Authentication The vehicle should be legitimate to participate to access smart parking navigation service. An attacker cannot impersonate a registered vehicle for enjoying free navigation service if the service is charged.
- Privacy Preservation The privacy of drivers would not be disclosed in smart parking navigation service. Moreover, given two parking queries, an attacker cannot learn whether both queries are delivered by the same driver, such that the driving pattern of the vehicle is protected.
- Traceability The TA recovers the real identities of the drivers participating in smart parking navigation service for service charging or unqualified driver identification.

3.2 P-SPAN

3.2.1 Overview of P-SPAN

P-SPAN has five stages: system setup, service registration, parking query, result retrieval, and driver tracing. We first provide a high-level description of P-SPAN, which is derived from the PS signature [19] and Bloom filters [20].

- **System Setup** TA bootstraps the navigation system by generating system parameters. The TA, the cloud and each fog initialize their individual secret-public key pairs, respectively. To obtain a valid digital driving licence, a driver interacts with the TA by executing the zero-knowledge proof-of-knowledge (ZKPoK) protocol based on the PS signature. The driver commits two values (w, w') and receives the TA's signature (A_1, A_2, A_3) on (w, w'), while the TA can learn nothing about (w, w'). The digital driving licence consists of the public part (ID, W, A_3) and the secret part $(w, \widehat{W}_0, A_1, A_2)$.
- **Service Registration** A driver registers the smart parking navigation service on the cloud by executing the ZKPoK protocol derived from the PS signature. The driver makes a commitment and authenticates the identifier W. After the execution of this protocol, the driver acquires the anonymous credential (B_1, B_2).
- **Parking Query** To discover a vacant parking space, the driver encrypts the basic query information and proves its identity by executing zero-knowledge proof protocol. The driver also randomizes (A_1, A_2, A_3) to obtain a group signature $(\widetilde{A}_1, \widetilde{A}_2, c, \tau)$ on the parking query. The query Q is forwarded to the cloud through the relay of the nearby fog. At last, the cloud knows the driver's destination and discovers an accessible parking lot for the vehicle.
- **Result Retrieval** The cloud protects a navigation result and replies it to the group of fogs $\mathscr{R}$ predicted to pass by for the querying driver. Each fog stores the navigation result on VBF_K to wait the driver to retrieve. If a driver enters a coverage area of a fog, it forwards a retrieving query, which contains a search index K^* and a group signature on K^* to preserve the driver's identity. The fog searches its VBF_K to retrieve the matched navigation result R if exists; and the driver decrypts to obtain the navigation result.
- **Driver Tracing** The TA is able to obtain the digital driving licence of the driver by opening the group signature $(\widetilde{A}_1, \widetilde{A}_2, c, \tau)$.

3.2.2 The Detailed P-SPAN

3.2.2.1 System Setup

The TA sets the security parameter ϱ, and $\varrho = 160$ or 256 generally for Elliptic Curve cryptography. Let p be a large prime with ϱ bits, and $(\mathbb{G}_1, \mathbb{G}_2, \mathbb{G}_T)$ be a set of cyclic groups with the same order p. $\hat{e} : \mathbb{G}_1 \times \mathbb{G}_2 \rightarrow \mathbb{G}_T$ is the type 3 bilinear pairing. g is a generator of $\mathbb{G}_1$ with $g \neq 1_{\mathbb{G}_1}$, and $\hat{g}$, $\hat{g}_0$ are two generators of $\mathbb{G}_2$ with $\hat{g} \neq \hat{g}_0 \neq 1_{\mathbb{G}_2}$. $\mathscr{H} : \{0, 1\}^* \rightarrow \mathbb{Z}_p$ is a collision-resistant secure hash function, $\mathscr{C} = ENC_{AES}(\mathscr{K}, \mathscr{M})$ and $\mathscr{M} = DEC_{AES}(\mathscr{K}, \mathscr{C})$ are the encryption and decryption algorithms of advanced encryption standard (AES), respectively. The TA randomly selects $(y, y_1) \in_R \mathbb{Z}_p^2$ and computes $\widehat{Y} = \hat{g}^y$, $\widehat{Y}_1 = \hat{g}^{y_1}$. (y, y_1) is TA's secret key and $(g, \hat{g}, \widehat{Y}, \widehat{Y}_1)$ is the corresponding public key.

The cloud initializes the parking space management service and smart parking navigation service. It manages the status information about parking spaces and utilizes the real-time parking information to provide smart parking navigation. To generate the secret-public key pair, the cloud randomly selects $(x, x_1, x_2, x_3) \in_R \mathbb{Z}_p^4$ to generate

$$(X, X_1, X_2, X_3) \leftarrow (g^x, g^{x_1}, g^{x_2}, g^{x_3}).$$

$$(\widehat{X}, \widehat{X}_1, \widehat{X}_2, \widehat{X}_3) \leftarrow (\hat{g}^x, \hat{g}^{x_1}, \hat{g}^{x_2}, \hat{g}^{x_3}).$$

where (x, x_1, x_2, x_3, X) is the cloud's secret key, and $(X_1, X_2, X_3, \widehat{X}, \widehat{X}_1, \widehat{X}_2, \widehat{X}_3)$ is the public key.

Each fog has a unique number RID associated with its location. The fog randomly chooses $z \in_R \mathbb{Z}_p$ as its secret key and generates $Z = g^z$ as its public key. The fog defines two Bloom filters: CBF_K and VBF_K. CBF_K is a (m, n, k, H, λ)-counting Bloom filter and VBF_K is a variant of the traditional Bloom filter. k hash functions $h_l \in H$ in both Bloom filters are defined as $h_l : \mathbb{G}_1 \rightarrow \mathbb{Z}_m$, for $1 \leq l \leq k$. VBF_K uses an array of γ-bit strings to denote the storage addresses of navigation results. A storage address S is divided into k shares of γ-bit, $S_1, S_2, \cdots, S_k$, based on the XOR-based secret sharing, and each share is stored on an index in VBF_K based on the hash values of the input. Initially, the counters in CBF_K and the strings in VBF_K are set to be zero.

A driver with a unique identity ID registers at the TA to receive a digital driving licence, after passing driving tests. The driver interacts with the TA in the following steps:

- The driver selects random values $(w, w') \in_R \mathbb{Z}_p^2$ to calculate $(W, \widehat{W}, \widehat{W}', \widehat{W}_0) \leftarrow (g^w, \widehat{Y}_1^w \widehat{Y}^{w'}, \hat{g}^{w'}, \widehat{Y}_1^w)$, and delivers $(ID, W, \widehat{W}, \widehat{W}')$ to the TA, along with the zero-knowledge proof:

$$\mathscr{PK}_1 = \{(w, w') : W = g^w \wedge \widehat{W} = \widehat{Y}_1^w \widehat{Y}^{w'} \wedge \widehat{W}' = \hat{g}^{w'}\}. \tag{3.1}$$

- The TA calculates $\widehat{W}_1 = \widehat{W}/\widehat{W}'^y$, verifies the proof $\mathscr{PK}_1$ and checks whether $\hat{e}(W, \widehat{Y}_1) = \hat{e}(g, \widehat{W}_1)$ holds. If either is invalid, the TA returns failure and aborts. Otherwise, the TA randomly chooses $v \in_R \mathbb{Z}_p$ to compute

$$(A_1, A_2, A_3) \leftarrow (g^v, (g^y W^{y_1})^v, \hat{e}(A_1, \widehat{Y}_1)). \tag{3.2}$$

Finally, the TA forwards (ID, A_1, A_2, A_3) to the driver via a secure channel and stores $(ID, W, \widehat{W}_1)$ in a database privately.

- The driver sets the digital driving licence, which has two parts, the public part (ID, W, A_3) and the secret part $(w, \widehat{W}_0, A_1, A_2)$. The secret part is kept secretly in a USB device and plugged in the vehicle when the driver starts the vehicle.

3.2.2.2 Service Registration

To enjoy parking navigation service, a vehicle registers on the cloud to obtain an anonymous credential for anonymous service access. OBU on the driver's vehicle interacts with the cloud in the following way:

- The OBU selects random values $(t, s) \in_R \mathbb{Z}_p^2$ to calculate $C = g^t X_1^{ID} X_2^s X_3^w$ and the zero-knowledge proof:

$$\mathscr{P}\mathscr{K}_2 = \{(t, s, w) : C = g^t X_1^{ID} X_2^s X_3^w \wedge W = g^w\}. \tag{3.3}$$

  The OBU forwards $(ID, C, W, \mathscr{P}\mathscr{K}_2)$ to the cloud.
- The cloud verifies the validity of $\mathscr{P}\mathscr{K}_2$ and returns failure and aborts if $\mathscr{P}\mathscr{K}_2$ is invalid; otherwise, it chooses a random value $u \in_R \mathbb{Z}_p$ to compute $(B_1, B_2) \leftarrow (g^u, (XC)^u)$. The cloud forwards (B_1, B_2) to the OBU via a secure channel and keeps (ID, C, W, B_1, B_2) in its database.
- The OBU verifies $\hat{e}(B_1, \widehat{X})\hat{e}(B_1, \hat{g}^t \widehat{X}_1^{ID} \widehat{X}_2^s \widehat{X}_3^w) \stackrel{?}{=} \hat{e}(B_2, \hat{g})$. If yes, the OBU calculates $B_3 = B_2/B_1^t$, and acquires the anonymous credential $AC = (B_1, B_3)$. Finally, it stores (AC, s) in the read-only memory of the OBU.

3.2.2.3 Parking Query

When a driver ID uses the parking navigation service, the OBU forwards a parking query to the cloud for vacant parking space discovery in the destination. Having the digital driving licence and the anonymous credential AC, the OBU conducts a parking query as follows:

- Generate the basic query information, including current location CL, current time t_1, the destination DS, acceptable price range AP, expected arrival time t_2 and expiration time t_3.
- Encrypt (DS, CL, AP, t_2, t_3) by randomly selecting $r \in_R \mathbb{Z}_p$, and calculating $c_1 = g^r$, $c_2 = \mathscr{H}(c_1, X_1^r)$, and $c_3 = ENC_{AES}(c_2, DS||\,CL||AP||t_2||t_3)$.
- Randomly choose $\kappa \in_R \mathbb{Z}_p$ to calculate a temporary session key $U = \hat{g}^\kappa$, $L = \mathscr{H}(ID, DS, AP, t_2, t_3)$ and a tag $T = \hat{g}^w \hat{g}_0^{Ls}$.
- Randomly choose $(\alpha, \beta) \in_R \mathbb{Z}_p^2$ to calculate $AC' = (B_1', B_3') = (B_1^\alpha, (B_3 B_1^\beta)^\alpha)$ and conduct a zero-knowledge proof as

$$\mathcal{SPK}\left\{\begin{array}{c}(ID, w, s, \kappa, \beta):\\ \hat{e}(B_1', \widehat{X}\hat{g}^{\beta})\hat{e}(B_1', \widehat{X}_1^{ID}\widehat{X}_2^{s}\widehat{X}_3^{w}) = \hat{e}(B_3', \hat{g})\\ \wedge U = \hat{g}^{\kappa}\\ \wedge T = \hat{g}^{w}\hat{g}_0^{Ls}\end{array}\right\}(N),$$

where N is a random identifier of the parking query.

- Randomly choose $(r', r'') \in_R \mathbb{Z}_p^2$ to randomise (A_1, A_2, A_3) by calculating

$$(\widetilde{A}_1, \widetilde{A}_2, \widetilde{A}_3) \leftarrow (A_1^{r'}, A_2^{r'}, A_3^{r'r''}), \tag{3.4}$$

compute $c = \mathcal{H}(\widetilde{A}_1, \widetilde{A}_2, \widetilde{A}_3, N, t_1, U, T, AC', \mathcal{SPK}, c_1, c_3)$, $\tau = r'' + cw$, and output $(\widetilde{A}_1, \widetilde{A}_2, c, \tau)$ as a signature.

Finally, the OBU keeps (U, κ) and forwards the parking query $Q = (N, t_1, c_1, c_3, U, T, AC', \mathcal{SPK}, \widetilde{A}_1, \widetilde{A}_2, c, \tau)$ to the nearby fog, if it is in the coverage area of a fog. Otherwise, the OBU forwards Q to the nearby vehicles to reach the nearby fog via V2V communications. The OBU also temporarily maintains the query Q and sends Q to a fog, when the vehicle connects the fog.

When a fog with RID receives Q from a vehicle, it first checks whether Q has been received and verifies the signature $(\widetilde{A}_1, \widetilde{A}_2, c, \tau)$ by computing $A = \hat{e}(\widetilde{A}_1, \widehat{Y}^{c})\hat{e}(\widetilde{A}_2, \hat{g}^{-c})\ \hat{e}(\widetilde{A}_1, \widehat{Y}_1^{\tau})$ and checking whether $c \stackrel{?}{=} \mathcal{H}(\widetilde{A}_1, \widetilde{A}_2, A, N, t_1, U, T, AC', \mathcal{SPK}, c_1, c_3)$ holds. If yes, the fog checks whether Q has the same tag T with a received query; otherwise, the fog returns failure. If the tag T is the same with the tag in a previous query, the fog ignores Q, otherwise, it chooses a random value $r_2 \in_R \mathbb{Z}_p$ and computes a signature on Q by executing $A_r = g^{r_2}$, $c_r = \mathcal{H}(RID, Q, A_r)$, $\tau_r = r_2 + zc_r$, and forwards (RID, Q, A_r, τ_r) to the cloud.

The cloud verifies the fog's signature by calculating $c_r' = \mathcal{H}(RID, Q, A_r)$ and checking $A_r Z^{c_r'} \stackrel{?}{=} g^{\tau_r}$ after receiving (RID, Q, A_r, τ_r). If not, the cloud returns failure; otherwise, it verifies whether T in Q is equal to the one in a received query. If yes, the cloud ignores this query; otherwise, it checks the signature $(\widetilde{A}_1, \widetilde{A}_2, c, \tau)$ to ensure the validity of ID' driving licence and $\mathcal{SPK}$ to authenticate the validity of the credential AC. If both are valid, the cloud decrypts (c_1, c_3) to obtain $DS||CL||AP||t_2||t_3$ as $c_2' = \mathcal{H}(c_1, c_1^{x_1})$, $DS||CL||AP||t_2||t_3 = DEC_{AES}(c_2', c_3)$. If the query has not expired, the cloud finds an available parking lot for the vehicle based on (DS, CL, AP, t_2) and the real-time parking data of parking lots. In addition, the cloud forwards Q to the TA for charging.

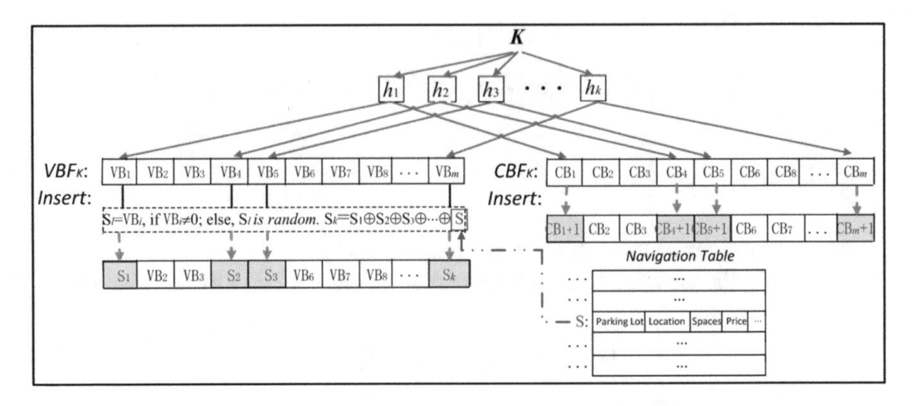

Fig. 3.2 Insert operation for the fog

3.2.2.4 Result Retrieval

The cloud generates a navigation result RS for the parking navigation query, including the geographic location of available parking lot, the parking price, the recommended parking space, and the quantity of vacant parking spots. The cloud also selects a random $\psi_1 \in_R \mathbb{Z}_p$ to generate $\phi_1 = g^{\psi_1}$, $\phi_2 = \mathscr{H}(\phi_1, U^{\psi_1})$, $\phi_3 = ENC_{AES}(\phi_2, RS)$ and $K = U^{x_1}$. Then, it computes a signature by selecting random $\psi_3 \in_R \mathbb{Z}_p$ and computes $\sigma_1 = g^{\psi_3}$, $\sigma_2 = \mathscr{H}(t_3, K, \phi_1, \phi_3, \sigma_1)$ and $\sigma_3 = \psi_3 + x_1\sigma_2$. After that, the cloud predicts the driving route of the vehicle and determines the set of fogs $\mathscr{R}$ that the vehicle would pass by. Finally, the cloud forwards the navigation result $R = (t_3, K, \phi_1, \phi_3, \sigma_1, \sigma_3)$ to each fog in $\mathscr{R}$. If the recommended parking space is possessed, the cloud generates a new navigation result R^* and forwards it to the fogs in $\mathscr{R}$ for the querying vehicle.

Upon receiving R, each fog in $\mathscr{R}$ computes $\sigma_2' = \mathscr{H}(t_3, K, \phi_1, \phi_3, \sigma_1)$ and verifies (σ_1, σ_3) as $\sigma_1 X_1^{\sigma_2'} \stackrel{?}{=} g^{\sigma_3}$. The fog returns failure if the equation inholds; otherwise, the fog stores R as shown in Fig. 3.2:

- Insert K into CBF_K. To be specific, the counter $CB_{h_l(K)}$ increases by one and the rest counters keep the same for each $1 \leq l \leq k$.
- Store R in the memory and obtain the storage address S.
- Insert S into VBF_K. Concretely, the fog divides S into k shares of γ-bit, $S_1, S_2, \cdots, S_k$, based on the XOR-based secret sharing. If the location on $h_l(K)$ of VBF_K has been occupied, the fog reuses the string $VB_{h_l(K)}$, i.e., S_l is set to be $VB_{h_l(K)}$, in which $l \in \{1, \cdots, k-1\}$; otherwise, S_l is fixed to be a random γ-bit string. The last string S_k is computed as $S_k = S \oplus S_1 \oplus S_2 \oplus \cdots \oplus S_{k-1}$, if $VB_{h_k(K)} = 0$; otherwise, find an unpossessed location on $h_l(K)$ to set $S_l = S \oplus S_1 \oplus \cdots \oplus S_{l-1} \oplus S_{l+1} \oplus \cdots \oplus S_k$.

When the vehicle enters the coverage area of a fog*, it firstly needs to ensure that the parking navigation result R is maintained on the fog* and retrieves the result from fog*. The OBU retrieves (U, κ) from its memory and computes $K^* = \widehat{X}_1^{\kappa}$. Then, the OBU chooses random $(u_1, u_2) \in_R \mathbb{Z}_p^2$ and computes a signature as

$$(C_1, C_2, C_3) \leftarrow (A_1^{u_1}, A_2^{u_1}, A_3^{u_1 u_2}),$$
$$\beta_1 = \mathscr{H}(C_1, C_2, C_3, K^*, \tilde{t}),$$
$$\tau_1 = u_2 + \beta_1 w,$$

where $\tilde{t}$ is the timestamp. At last, the OBU sends the retrieving query $(K^*, C_1, C_2, \beta_1, \tau_1, \tilde{t})$ to the fog* for navigation result retrieval.

Upon receiving the retrieving query from the OBU, the fog* performs the following steps to find the corresponding navigation result.

- Verify the signature $(C_1, C_2, \beta_1, \tau_1)$ by computing $C_3' = \hat{e}(C_1, \widehat{Y}^{\beta_1})\hat{e}(C_2, \hat{g}^{-\beta_1})$ $\hat{e}(C_1, \widehat{Y}_1^{\tau_1})$ and verifying whether $\beta_1 = \mathscr{H}(C_1, C_2, C_3', K^*, \tilde{t})$ or not, and return failure and abort if the equation does not hold.
- Check whether the counters on the locations $(h_1(K^*), \cdots h_k(K^*))$ in CBF_K are nonzero, and return failure and abort if one of the counters is zero.
- Obtain the storage address S as $S = VB_{h_1(K^*)} \oplus VB_{h_2(K^*)} \oplus \cdots \oplus VB_{h_k(K^*)}$ and search the navigation result R on the storage address S.
- Pick a random value $r_3 \in_R \mathbb{Z}_p$ and calculate $\sigma_1^* = g^{r_3}$, $\sigma_2^* = \mathscr{H}(RID^*, R, \sigma_1^*)$ and $\sigma_3^* = r_3 + z^* \sigma_2^*$.

The fog* forwards $(RID^*, R, \sigma_1^*, \sigma_3^*)$ to the OBU and performs the deletion operation to remove K^* from CBF_K and S from VBF_K. Specifically, the counters in CBF_K on the indices $h_l(K^*)$ for $1 \le l \le k$ decrease by one, and the shares of S in VBF_K are removed if the corresponding counters in CBF_K are set to be zero. In addition, if the navigation result is expired, the fog performs deletion operation to update CBF_K and VBF_K.

Upon receiving $(RID^*, R, \sigma_1^*, \sigma_3^*)$, the OBU calculates $\sigma_4^* = \mathscr{H}(RID^*, R, \sigma_1^*)$ and verifies whether $\sigma_1^*(Z^*)^{\sigma_4^*} = g^{\sigma_3^*}$. If not, the OBU returns failure; otherwise, it calculates $\sigma_4 = \mathscr{H}(t_3, K, \phi_1, \phi_3, \sigma_1)$ and checks whether $\sigma_1 X_1^{\sigma_4} = g^{\sigma_3}$. If not, the OBU delivers R to the TA for complaint; otherwise, the OBU obtains $\phi_2' = \mathscr{H}(\phi_1, \phi_1^{\kappa})$ and recovers the navigation result $RS = DEC_{AES}(\phi_2', \phi_3)$. According to navigation result, the driver finds the accessible parking space near the destination. When the vehicle drives through other fogs, it would till forward retrieving queries to the nearby fogs to detect whether the result is updated and retrieve the latest one.

3.2.2.5 Driver Tracing

The TA can know the digital driving licence of the driver from the signature $(\widetilde{A}_1, \widetilde{A}_2, c, \tau)$. To be specific, the TA utilizes each $(ID, W, \widehat{W}_1)$ to test whether $\hat{e}(\widetilde{A}_2, \hat{g}) = \hat{e}(\widetilde{A}_1, \widehat{Y})\, \hat{e}(\widetilde{A}_1, \widehat{W}_1)$ holds or not, until it gets a match.

3.3 Security Discussion

We discuss the security properties of our proposed P-SPAN, including identity authentication, service authentication, privacy preservation and traceability.

Identity Authentication: The driver with a qualified driving licence is eligible to drive on roads. The driving licence is generated based on the PS signature interacting with the TA, who is responsible for issuing driving licences to drivers. The TA releases the driving licence $(ID, W, w, \widehat{W}_0, A_1, A_2, A_3)$ to the driver in the system setup phase, which is randomized to compute the signature on the parking query $(\widetilde{A}_1, \widetilde{A}_2, c, \tau)$. In the signature verification, the verifier can learn whether the driver is eligible for driving or not. Since the signature is generated from the driving licence $(ID, W, w, \widehat{W}_0, A_1, A_2, A_3)$, only the driver with an eligible driving licence can compute a valid signature $(\widetilde{A}_1, \widetilde{A}_2, c, \tau)$. Thus, the identity authentication of drivers relies on the driving licence, which is a PS signature generated by the TA. Since the PS signature is unforgeable based on the modified LRSW assumption 2, there is no attacker that can forge the driving licence $(ID, W, w, \widehat{W}_0, A_1, A_2, A_3)$ and further obtain the valid signatures on navigation queries.

Service Authentication: In service registration, the OBU interacts with the cloud to generate AC for smart parking navigation service access. To query a vacant parking space, the OBU proves the possession of AC using zero-knowledge proof to show the access capability of parking navigation service. Only the vehicles with valid anonymous credentials can enjoy this service. To generate AC, the cloud signs the commitment C with its secret key to compute a signature (B_1, B_2) and the vehicle obtains $AC = (B_1, B_3)$ from (B_1, B_2). The unforgeability of anonymous credential (B_1, B_3) is reduced to the modified LRSW assumption 1 [19]. AC satisfies $B_1 = g^u$, $B_3 = (Xg^t X_1^{ID} X_2^s X_3^w)^u / g^{ut} = (X X_1^{ID} X_2^s X_3^w)^u$, which is a valid PS signature on (ID, s, w). Therefore, the security of anonymous credential is reduced to the modified LRSW assumption 1, while the unforgeability of PS signature relies on the modified LRSW assumption 2 [19]. In summary, if the modified LRSW assumption 1 holds, any attacker cannot forge an anonymous credential for smart parking navigation service access.

Privacy Preservation: We prove that the driver's identity would not be disclosed in parking query and result retrieval phases. Firstly, in parking query phase, the driver delivers a parking navigation query $Q = (N, t_1, c_1, c_3, U, T, AC', \mathcal{SPK}, \widetilde{A}_1, \widetilde{A}_2, c, \tau)$ to the cloud, in which $(AC', \mathcal{SPK})$, $(\widetilde{A}_1, \widetilde{A}_2, c, \tau)$ and T are associated with the driver's identity. $(AC', \mathcal{SPK})$ would not leak the driver's identity as $AC' = (B_1', B_3')$ is randomized from (B_1, B_3) using (α, β) and the zero-knowledge proof $\mathcal{SPK}$ is sound. $(\widetilde{A}_1, \widetilde{A}_2, c, \tau)$ would not disclose any privacy about drivers, after (A_1, A_2, A_3) are randomized by random values (r', r'') and only the public key of TA is used to check the validity of the signature $(\widetilde{A}_1, \widetilde{A}_2, c, \tau)$. Although the tag T has the driver's secret value w, it is difficult for an attacker to identify the driver's identity or link two tags to the same driver, unless the Decisional Diffie-Hellman problem (DDH) in $\mathbb{G}_2$ [21] is easy to address. We claim that if $\mathscr{A}$ can identify an honest driver out of two challenging identities, there exists a simulator $\mathscr{S}$ to solve an instance of the DDH problem in $\mathbb{G}_2$, that is, given $(H, H_1, H_2, H_3) \in \mathbb{G}_2^4$, $\mathscr{S}$ can say whether there exists (ω_1, ω_2), such that $H_1 = H^{\omega_1}$, $H_2 = H^{\omega_2}$, $H_3 = H^{\omega_1 \omega_2}$. The security model of the identity privacy preservation is defined in [22] for the formalization of the adversary's capacity and the anonymity goal.

$\mathscr{S}$ setups the system parameters and sets $\hat{g} = H$, $\hat{g}_0 = H_1$. $\mathscr{S}$ picks two drivers' identities (ID_0, g^{w_0}) and (ID_1, g^{w_1}), in which $(w_0, w_1) \in_R \mathbb{Z}_p^2$ and give them to $\mathscr{A}$. $\mathscr{S}$ simulates the system setup and service registration on behalf of the TA and the cloud. $\mathscr{S}$ also interacts with $\mathscr{A}$ on behalf of the drivers ID_0 and ID_1 in the following interactions.

$\mathscr{S}$ honestly responds parking queries acting as ID_0. For ID_1, $\mathscr{S}$ randomly chooses $(\kappa, w, s, t_1, L) \in_R \mathbb{Z}_p^5$ to compute $U = H^{\kappa}$, $T = H^w H_1^{Ls}$, generates $(c_1, c_3, AC', \widetilde{A}_1, \widetilde{A}_2, c, \tau)$, and simulates the zero-knowledge proof $\mathscr{SPK}$ to interact with $\mathscr{A}$.

$\mathscr{S}$ picks $\beta \in \{0, 1\}$. If $\beta = 0$, $\mathscr{S}$ honestly generates a parking query; otherwise, $\mathscr{S}$ chooses random $(\kappa^*, w^*, t_1^*, L^*) \in_R \mathbb{Z}_p^4$ to generate $U^* = H^{\kappa^*}$, $T^* = H^{w^*} H_3^{L^*}$, and compute $(c_1^*, c_3^*, AC^*, \widetilde{A}_1^*, \widetilde{A}_2^*, c^*, \tau^*)$. $\mathscr{S}$ simulates the zero-knowledge proof $\mathscr{SPK}^*$ and sends them to $\mathscr{A}$. It is obvious that the game is perfectly simulated by $\mathscr{S}$ if $log_H H_3 = log_H H_1 \cdot log_H H_2$. Otherwise, $\mathscr{S}$ cannot contain any information about ID_0 and ID_1.

Finally, $\mathscr{A}$ returns β'. If $\beta' = \beta$, $\mathscr{S}$ confirms that there exists (ω_1, ω_2), such that $H_1 = H^{\omega_1}$, $H_2 = H^{\omega_2}$, $H_3 = H^{\omega_1 \omega_2}$. Therefore, $\mathscr{S}$ addresses the DDH problem in $\mathbb{G}_2$.

Secondly, in result retrieving, the driver's identity is preserved, since the retrieving query $(K^*, C_1, C_2, \beta_1, \tau_1, \tilde{t})$ would not disclose any privacy about the driver. K^* is a result of Diffie-Hellman agreement, which can be viewed as a random value, and $(C_1, C_2, \beta_1, \tau_1)$ is a signature randomized from (A_1, A_2, A_3) and only the TA' public key is needed for verification. Therefore, the driver's identity is not leaked in result retrieval.

Traceability: To recover the driver's identity, the TA utilize the stored $(ID, W, \widehat{W}_1)$ to verify $\hat{e}(\widetilde{A}_2, \hat{g}) = \hat{e}(\widetilde{A}_1, \widehat{Y})\hat{e}(\widetilde{A}_1, \widehat{W}_1)$, until finding a matched $(ID, W, \widehat{W}_1)$. Since $\widehat{W}_1$ is stored by the TA, only the TA can recover the vehicle's identity from $(\widetilde{A}_1, \widetilde{A}_2, c, \tau)$.

In summary, P-SPAN achieves identity authentication, service authentication, privacy preservation and traceability.

3.4 Retrieving Probability and Performance Analysis

We discuss the retrieving probability of navigation results for drivers and evaluate the computational and communication overhead of P-SPAN.

3.4.1 Retrieving Probability

To ensure the retrievability of navigation results, a counting Bloom filter CBF_K is used to count the number of collisions occur on each index, and a variant of Bloom filter VBF_K is designed to store the storage addresses of navigation results on the fog. Due to the false positive probability of a Bloom filter, drivers probably retrieve false navigation results, indicating that K^* does not exist in CBF_K, but all $CB_{h_l(K^*)}$ are non-zero, for $1 \leq l \leq k$. In CBF_K, the probability that a counter is non-zero is $P = 1 - (1 - \frac{1}{m})^{kn}$. Thereby, the upper bound of false positive probability is

$$\epsilon = (1 + O(\frac{k}{P}\sqrt{\frac{\text{In}m - k\text{In}P}{m}}))P^k, \tag{3.5}$$

which is negligible in k. In VBF_K, the false positive probability is ϵ as well, as the shares of storage address S replace the counters in CBF_K from the high level point of view. Thus, the lower bound of retrieving probability is $1 - \epsilon$. In reality, the probability to retrieve a false navigation result is much smaller than ϵ, since the shares on the indices $h_1(K^*), \cdots, h_k(K^*)$ in VBF_K may not consist of a correct storage address, if K^* is not an element in CBF_K. If we require the probability of successful retrieval to be at least θ, the lower bound of m is $m > nlog_2e \cdot log_2 1/(1 - \theta)$, where e is the base of national logarithms.

The above result shows that a driver can obtain the correct navigation result from a fog, if the required result is stored on the fog. Now we assume the probability that the j-th driving-through fog is storing the navigation result, when the driver sends its retrieving query, is ϕ_j. The first fog is the one that sends the parking query, and the driver would drive through ν fogs before it arrives its desired destination. Thereby, the probability that the driver successfully acquires the navigation result from the j-th fog is

$$\prod_{i=1}^{j-1}(1 - \phi_i)(1 - \theta)^{j-1}\phi_j, \tag{3.6}$$

and thereby the probability that the driver obtains the navigation result, before it arrives the destination, is

$$\sum_{j=1}^{\nu}(\prod_{i=1}^{j-1}(1 - \phi_i)(1 - \theta)^{j-1}\phi_j). \tag{3.7}$$

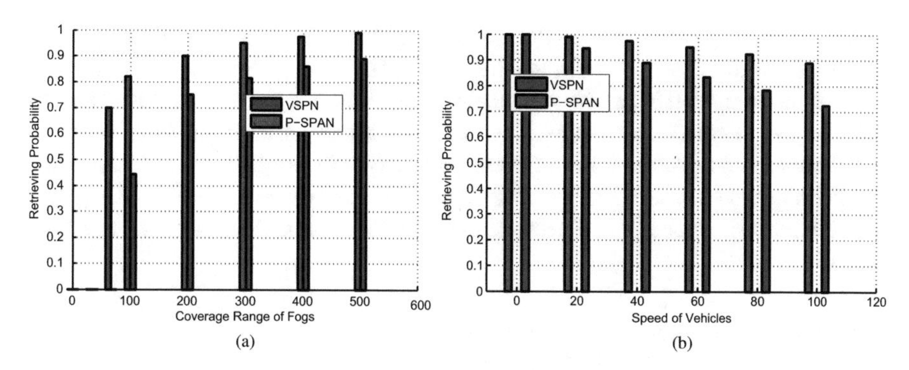

Fig. 3.3 Comparison on retrieving probability. (**a**) The range of fogs. (**b**) The speed of vehicles

We calculate the retrieving probability of navigation results. Based on the testing result in [23], a vehicle needs about 1s to build a connection with a fog if the packet lost probability is 50%, due to the poor channel condition, and 15 vehicles in the queue to access the fog. The maximum coverage range of a fog is 500 meters, and the delay of smart parking navigation query is around 2s considering the queueing delay, transmission delay in two-round interactions and the processing delay for the cloud. The OBU conducts parking queries, forwards to the nearby vehicles or fogs at anytime, and retrieves to the navigation results in VSPN. In P-SPAN, the OBU receives the results from the querying fogs if the connection is active; otherwise, it retrieves the results from new fogs when driving through. The simulation result is shown in Fig. 3.3. With the increasing fogs' coverage range, the retrieving probability of navigation results increase significantly for both VSPN and P-SPAN. Our P-SPAN has higher retrieving probability than VSPN in Fig. 3.3a, as the OBUs can retrieve the navigation results from the passing-by fogs. If the speed of vehicle increases, the retrieving probability would be decreased, but P-SPAN still has higher retrieving probability compared with VSPN, as shown in Fig. 3.3b.

3.4.2 Computational Overhead

We evaluate the computational overhead of P-SPAN by counting the number of time-consuming cryptographic operations in each phase, including scalar multiplication in $\mathbb{G}_1/\mathbb{G}_2$, AES encryption/decryption, exponentiation in $\mathbb{G}_T$ and bilinear pairing. Other operations, such as point addition, integer multiplication and hash function, are not resource-consuming compared with scalar multiplication and bilinear pairing. We use T_{SM}, T_{AES}, T_{Exp} and T_p to denote the running time of scalar multiplication in $\mathbb{G}_1/\mathbb{G}_2$, AES encryption/decryption, exponentiation in $\mathbb{G}_T$ and bilinear pairing for vehicles, respectively. To demonstrate the high efficiency of our P-SPAN, we compare P-SPAN with VSPN [15] and show the comparison results in Table 3.1. Although P-SPAN is less efficient than VSPN in service registration phase, this phase is executed only once for each vehicle. In parking query phase, four bilinear pairings $\hat{e}(B_1, \widehat{X})$, $\hat{e}(B_1, \widehat{X}_1)$, $\hat{e}(B_1, \widehat{X}_2)$ and $\hat{e}(B_1, \widehat{X}_3)$ can be pre-computed in service registration phase with the aid of the cloud. Thus, no bilinear pairing is executed in parking query and result retrieval phases of P-SPAN. Furthermore, P-SPAN is more efficient than VSPN in result retrieval phase, since VSPN requires each OBU to perform 4ν bilinear pairings to retrieve the navigation result from fogs, where ν is the number of fogs to generate the navigation result for the driver.

Table 3.1 Computational overhead of vehicles

Phases	VSPN	P-SPAN
System setup	$3T_{SM} + T_p + T_{AES}$	$8T_{SM}$
Vehicle registration	$6T_{SM} + T_{AES}$	$13T_{SM} + 3T_p$
Parking query	$T_{SM} + T_{AES}$	$14T_{SM} + 4(T_p) + T_{AES}$ + $4T_{Exp}$
Result retrieval	$4\nu T_p$	$9T_{SM}$

P-SPAN is a VANET-based smart parking navigation system implementable on OBUs, fogs and the cloud, which brings huge convenience to drivers on parking space discovery. To evaluate the practicality of P-SPAN, we execute our P-SPAN on a notebook with Intel Core i5-4200U CPU @2.29GHz and 4.00GB memory. We use MIRACL library 5.6.1 [24] to implement number-theoretic based methods of cryptography. The R-ATE pairing [25] is utilized to realize the bilinear pairing. To ensure the security of P-SPAN, the parameter p is approximately 160 bits. The execution time of OBU in system setup and service registration phases is 30.376 ms and 143.649 ms, respectively. The OBU had to execute approximately 62.535 ms and 38.284 ms to deliver a parking navigation query and retrieve the navigation result. Therefore, P-SPAN is computation-efficient to be implemented on OBUs. Fig. 3.4a shows the comparison result between P-SPAN and VSPN about the time cost of OBU in result retrieval phase. The computational overhead of OBU in result retrieval phase of P-SPAN is constant and pretty low, while the executing time of OBU to read the navigation result in VSPN is linear with the number of fogs participating in the navigation result generation.

3.4.3 Communication Overhead

To demonstrate the communication overhead of P-SPAN, we count the length of exchanged messages among vehicles, fogs and the cloud. The system parameters are set to be the same as those in the simulation, in which $\varrho = 160$. In each parking query phase, the OBU needs to deliver a smart parking navigation query Q to the nearby fog, which is $5216 + |N| + |DS| + |CL| + |AP| +$

$|t_1| + |t_2| + |t_3|$ bits, where $|N|, |DS|, |CL|, |AP|, |t_1|, |t_2|, |t_3|$ are the binary length of $N, DS, CL, AP, t_1, t_2, t_3$, respectively. The fog verifies the signature $(\widetilde{A}_1, \widetilde{A}_2, c, \tau)$, appends a 672-bit Schnorr signature (A_r, τ_r) to Q and forwards (A_r, τ_r, Q) to the cloud. Upon receiving the parking query, the cloud generates the navigation result R with $1696 + |t_3| + |RS|$ bits and sends R to the fogs that the querying vehicle may drive through, where $|RS|$ is the binary length of RS. When the vehicle enters the coverage area of a fog*, it sends $1856+|\tilde{t}|$-bit $(K^*, C_1, C_2, \beta_1, \tau_1)$ to the fog*, where $|\tilde{t}|$ is the binary length of $\tilde{t}$. If R is maintained on fog*, fog* returns $(RID^*, R, \sigma_1^*, \sigma_3^*)$ to the querying vehicle, which is $2368 + |RID^*| + |t_3| + |RS|$ bits, where $|RID^*|$ denotes the binary length of RID^*.

We compare the communication overhead of P-SPAN and VSPN in result retrieval phase in Fig. 3.4b. We assume the length of navigation result RS in P-SPAN is equal to that in VSPN and $|RID^*| = |t_3| = 160$ bits. The communication overhead of the OBU is constant in our P-SPAN, while the overhead increases linearly with respect to the number of fogs participating in navigation result generation in VSPN.

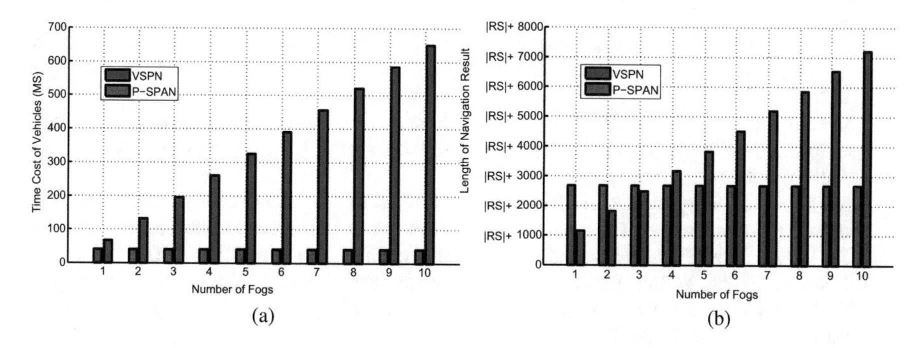

Fig. 3.4 Performance comparison for OBUs on result retrieval. (**a**) Time cost. (**b**) Communication cost

3.5 Summary

In this chapter, we have developed a privacy-enhanced smart parking navigation scheme based on Bloom filter and vehicular communications. In P-SPAN, a driver can query a vacant parking space to a cloud and retrieve the navigation results without exposing identity privacy. An efficient data retrieving mechanism is also proposed to improve the retrieving probability of navigation results for anonymous vehicular communications under assumption that it is difficult for vehicles to hold the connections with the fogs due to the high mobility. In addition, we have extended the Bloom filter to reduce storage overhead and collusion probability for fogs. Finally, we have demonstrated that P-SPAN reaches the desirable security and privacy goals, and shown its efficiency and practicality for implementation in performance evaluation.

References

1. R. Lu, X. Lin, H. Zhu, and X. Shen, "An intelligent secure and privacy-preserving parking scheme through vehicular communications," *IEEE Transactions on Vehicular Technology*, vol. 59, no. 6, pp. 2772–2785, 2010.
2. H. Li, M. Dong, and K. Ota, "Control plane optimization in software-defined vehicular ad hoc networks," *IEEE Transactions on Vehicular Technology*, vol. 65, no. 10, pp. 7895–7904, 2016.
3. J. Ni, X. Lin, K. Zhang, and X. Shen, "Privacy-preserving real-time navigation system using vehicular crowdsourcing," in *Proc. of VTC-Fall*, 2016, pp. 1–5.
4. P. R. Pereira, A. Casaca, J. J. Rodrigues, V. N. Soares, J. Triay, and C. Cervello-Pastor, "From delay-tolerant networks to vehicular delay-tolerant networks," *IEEE Communications Surveys & Tutorials*, vol. 14, no. 4, pp. 1166–1182, 2012.
5. R. Lu, X. Lin, H. Zhu, P.-H. Ho, and X. Shen, "Ecpp: Efficient conditional privacy preservation protocol for secure vehicular communications," in *Proc. of IEEE INFOCOM*, 2008, pp. 1229–1237.
6. K. Zhang, J. Ni, K. Yang, X. Liang, J. Ren, and X. Shen, "Security and privacy in smart city applications: Challenges and solutions," *IEEE Communications Magazine*, vol. 55, no. 1, pp. 122–129, 2017.
7. L. Guo, M. Dong, K. Ota, Q. Li, T. Ye, J. Wu, and J. Li, "A secure mechanism for big data collection in large scale internet of vehicle," *IEEE Internet of Things Journal*, vol. 4, no. 2, pp. 601–610, 2017.
8. M. A. Pandi Vijayakumar, A. Kannan, and L. J. Deborah, "Dual authentication and key management techniques for secure data transmission in vehicular ad hoc networks," *IEEE Transactions on Intelligent Transportation Systems*, vol. 17, no. 4, pp. 1015–1028, 2016.
9. J. Sun, C. Zhang, Y. Zhang, and Y. Fang, "An identity-based security system for user privacy in vehicular ad hoc networks," *IEEE Transactions on Parallel and Distributed Systems*, vol. 21, no. 9, pp. 1227–1239, 2010.
10. J. Cheng, "How apple tracks your location without consent, and why it matters," *Ars Technica*, 2011.
11. D. J. Wu, J. Zimmerman, J. Planul, and J. C. Mitchell, "Privacy-preserving shortest path computation," *arXiv preprint arXiv:1601.02281*, 2016.
12. D. Liu, J. Ni, X. Lin, and X. Shen, "Anonymous group message authentication protocol for lte-based v2x communications," *Internet Technology Letters*, vol. 1, no. 2, p. e25, 2018.
13. M. D. Dikaiakos, A. Florides, T. Nadeem, and L. Iftode, "Location-aware services over vehicular ad-hoc networks using car-to-car communication," *IEEE Journal on Selected Areas in Communications*, vol. 25, no. 8, 2007.
14. J.-L. Huang, L.-Y. Yeh, and H.-Y. Chien, "Abaka: An anonymous batch authenticated and key agreement scheme for value-added services in vehicular ad hoc networks," *IEEE Transactions on Vehicular Technology*, vol. 60, no. 1, pp. 248–262, 2011.
15. T. W. Chim, S.-M. Yiu, L. C. Hui, and V. O. Li, "Vspn: Vanet-based secure and privacy-preserving navigation," *IEEE Transactions on Computers*, vol. 63, no. 2, pp. 510–524, 2014.
16. Y. He, J. Ni, X. Wang, B. Niu, F. Li, and X. Shen, "Privacy-preserving partner selection for ride-sharing services," *IEEE Transactions on Vehicular Technology*, vol. 67, no. 7, pp. 5994–6005, 2018.
17. Q. Jiang, J. Ni, J. Ma, L. Yang, and X. Shen, "Integrated authentication and key agreement framework for vehicular cloud computing," *IEEE Network*, vol. 32, no. 3, pp. 28–35, 2018.

18. J. Ni, K. Zhang, X. Lin, Y. Yu, and X. Shen, "Cloud-based privacy-preserving parking navigation through vehicular communications," in *Proc. of Securecomm*, 2016, pp. 85–103.
19. D. Pointcheval and O. Sanders, "Short randomizable signatures," in *Proc. of CT-RSA*, 2016, pp. 111–126.
20. C. Dong, L. Chen, and Z. Wen, "When private set intersection meets big data: an efficient and scalable protocol," in *Proc. of ACM CCS*, 2013, pp. 789–800.
21. D. Boneh and X. Boyen, "Short signatures without random oracles," in *Proc. of EUROCRYPT*, 2004, pp. 56–73.
22. M. H. Au, J. K. Liu, J. Fang, Z. L. Jiang, W. Susilo, and J. Zhou, "A new payment system for enhancing location privacy of electric vehicles," *IEEE transactions on vehicular technology*, vol. 63, no. 1, pp. 3–18, 2014.
23. W. Xu, H. A. Omar, W. Zhuang, and X. Shen, "Delay analysis of in-vehicle internet access via on-road wifi access points," *IEEE access*, vol. 5, pp. 2736–2746, 2017.
24. M. Scott, "Multiprecision integer and rational arithmetic c/c++ library (miracl)," *URL:* http://www.shamus.ie, 2003.
25. J.-L. Beuchat, J. E. González-Díaz, S. Mitsunari, E. Okamoto, F. Rodríguez-Henríquez, and T. Teruya, "High-speed software implementation of the optimal ate pairing over barreto–naehrig curves," in *Proc. of Pairing*, 2010, pp. 21–39.

Chapter 4
Location Privacy Protection in Mobile Crowdsensing

With the increasingly popularity of user-centric mobile sensing and computing devices, e.g., smart phones, in-vehicle sensing devices and wearable devices, our knowledge of the physical world is extended by opening a new door to collect and process data about social events and natural phenomena [1, 2]. This alternative has triggered the emergence of mobile crowdsensing (MCS) services [3]. In MCS, individuals cooperatively sense data for the tasks released by customers and extract information to measure and map phenomena of common interests using their mobile devices [4].

In addition, MCS triggers a series of challenges towards customers and mobile users, e.g., privacy leakage, which means that MCS puts the privacy of both mobile users and customers at risk [5–8]. On one hand, the collected data by mobile devices may inadvertently reveal sensitive information about mobile users, such as location, preferences, daily routines, habits, health status, and political affiliation. Moreover, the more tasks mobile users participate in and the richer the data they report, the higher the risk that their personal information will be exposed. For instance, if a mobile user submits a medical experience to a department, the service provider will be aware that the mobile user may have a disease related with the department. Thus, protecting the privacy of mobile users is the first-order security concern in MCS. Without privacy preserving mechanisms, mobile users may not be willing to engage in MCS activities [9]. On the other hand, a releasing task may pose sensitive data about the customer, including, purchase intention and points of interest [10]. For example, a house agency may learn Bob desires to purchase a house in a specific area, if he collects the information about traffic condition and noise level in this area. Several privacy-preserving MCS schemes [11, 12] have been introduced to prevent privacy invasion for either mobile users or customers using anonymity techniques [13]. Unfortunately, anonymization is not enough for privacy preservation, since sensitive data can be disclosed to nearby eavesdroppers via physical observation. In addition, the de-anonymity technique has matured nowadays [13]. For instance, the sensing data is always cross-referenced with other data, e.g., social graph and

X. Lin et al., *Privacy-Enhancing Fog Computing and Its Applications*,
SpringerBriefs in Electrical and Computer Engineering,
https://doi.org/10.1007/978-3-030-02113-9_4

mobility patterns, which end up with the fact that adversaries might identify the data source even if the identity is invisible. Human mobility patterns can be predicted and an anonymous user can be identified from four data points [14]. An anonymous sensing report may be easily linked to a particular user by analyzing some public or available data over the Internet, particularly in the big data era. In short, it is critical to deal with the privacy problems in MCS.

Once the sensitive data of customers and mobile users is perfectly protected, a new obstacle on task allocation for the service provider is observed [15]. Concretely, the group of mobile users always affects the quality of sensing results. An improper task allocation policy may trigger plenty of troubles for the mobile users, particularly in location-based applications, as the engaging mobile users need to stay in the sensing points to sense data [16]. We aim to optimize task allocation for location-based applications, which leverages spatial and temporal correlation of mobile users [17]. However, it leaks the points of interest of customers and the location of mobile users to the service provider, by which the anonymity of both customers and mobile users is invaded. Thus, it is challenging to effectively allocate crowdsensing tasks without privacy leakage.

Fog computing [18] is a new paradigm that offers computing, storage and networking services between terminal devices and the Internet with appealing properties, including location awareness, geographic distribution and low response latency [18]. With fog computing, a huge number of decentralized mobile devices can self-organize to communicate and potentially collaborate with each other via a fog node located at the edge of the Internet. Inspired by fog computing, in this chapter, we introduce a Privacy-Preserving Fog-assisted Mobile Crowdsensing framework (PPFMC) for location-based applications [19]. The main contributions are two-fold.

- A fog-based task allocation method is designed without exposing location information to the service provider, preserving conditional privacy for customers and mobile users. By using extended proxy re-encryption [20] and BBS+ signature [21], the registered customers and mobile users are allowed to anonymously prove their capacity to engage in the crowdsensing services and securely perform the crowdsensing tasks without disclosing tasks and sensing reports. Besides, to prevent mobile users from misbehaving for unfair reward, a trusted authority can detect greedy mobile users and trace their identities.
- A fog-based location matching approach is proposed to enable the fog to allocate the crowdsensing tasks according to the sensing areas of the tasks and the locations of mobile users. In specific, the customer and the mobile user utilize a self-defined random matrix to protect the sensing area and the location, respectively. Therefore, the fog determines whether the mobile user is located in the sensing area of the task without exposing the points of interest of customer and the location of the mobile user.

4.1 Problem Statement

We formally define the system model, threat model, and identify our design goals.

4.1.1 System Model

The system model of a typical fog-based MCS application (e.g., Gigwalk, mCrowd) consists of five entities: an authority, a service provider, fogs, customers and mobile users, as shown in Fig. 4.1. The fog-based MCS is performed in the following steps.

Step 1 Each customer or mobile user registers at a trusted authority.
Step 2 A customer creates a sensing task ST for data collection in a specific area L, and forwards it to the service provider, along with the authentication message.
Step 3 The service provider provides MCS services and releases the received task for the customer.
Step 4 The mobile users $U_{i\{i \in R\}}$ obtain their current or future location information via mobile devices that are capable of localization (e.g., by wireless access points or GPS) and forward their locations to the fogs, along with the authentication information.
Step 5 The fog reads the sensing area of the task L and checks whether the mobile users are in the area L or they will be in the area L in the near future.
Step 6 Suppose the locations of the mobile users $U_{i\{i \in \mathscr{L}\}}$ are consistent with the sensing area L. The fog distributes the task ST to the mobile users $U_{i\{i \in \mathscr{L}\}}$. Otherwise, the fog returns failure to the service provider.
Step 7 The mobile users accept or reject the task based on rewards and costs of performing the task. The mobile users, who accept the task, collect data for the event or phenomena, and generate sensing reports.
Step 8 The mobile users return the sensing reports to the service provider.
Step 9 The service provider forwards the sensing reports to the customer.
Step 10 The customer reads the reports and distributes rewards to the mobile users according to their contributions on the task.

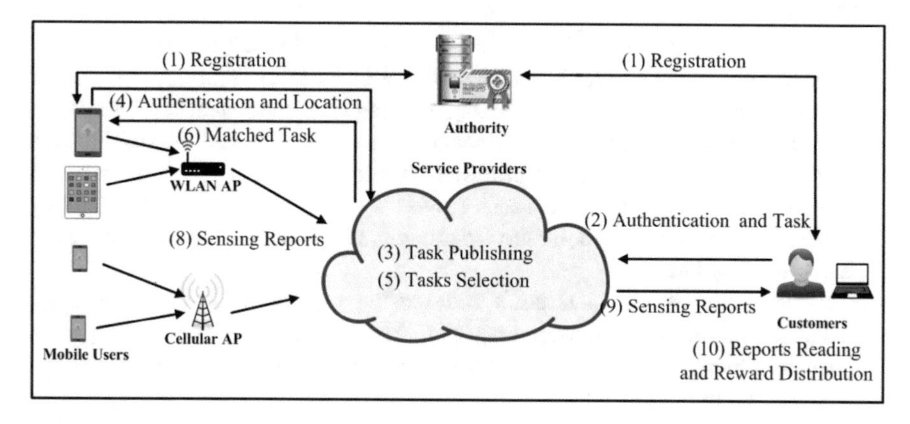

Fig. 4.1 System model of mobile crowdsensing

4.1.2 Threat Model

In threat model, the service provider honestly performs the MCS services, but is curious about the privacy of customers and mobile users. The information obtained by the service provider may contain plenty of knowledge, such as the identities of customers and mobile users, the places where the mobile users visit, or the area or events the customers are interested in. The service provider builds a spatia-temporal probability distribution for a specific mobile user and learns certain sensitive data about customers and mobile users, e.g., preference, political affiliation, social relation, and purchase intention. The fogs are also semi-honest. They are interested in the locations or trajectories of mobile users and the contents submitted the service provider. Although outside attackers may capture and exploit the sensitive data, the service provider is assumed not compromised to absurdly perform the operations, as the intrusion detection systems (IDS)/intrusion prevention systems (IPS) and Firewall are installed on the servers. Besides, the service provider would not collude with mobile users to invade the privacy of customers or other mobile users.

Mobile users are curious about the privacy about the customers and the other mobile users. Particularly, they are willing to have the other mobile users who are engaging in the same tasks to acquire more rewards, and know more data about customers to satisfy the requirements of customers. The location is obtained from the GPS chip in the mobile device or the access points, mobile users cannot modify their location information. Greedy mobile users may gain unfair award by double reporting the data in a time period, or collude together to send false reports to deceive the customer. However, the majority of mobile users are assumed to be honest.

4.1.3 Design Goals

To achieve privacy-preserving MCS to resist the security threats, our system should realize following objectives.

- Fog-based task allocation The releasing tasks are allocated to the fogs in the sensing area and then the fogs distribute the tasks to the mobile users. Other mobile users who are out of the sensing area can learn nothing about the tasks.
- Location privacy preservation The location of mobile users and the sensing area of releasing tasks needs to be protected. The mobile users are only aware of whether their geographical positions are in the sensing area or not. The sensing area of a task is invisible to the service provider and the mobile users who are not in the sensing area.
- Confidentiality of releasing tasks and sensing reports No entity, except the delegated mobile users (customers), can obtain the content of releasing tasks (sensing reports), such that the privacy of mobile users (customers) would not be disclosed to others.

- Anonymity of mobile users and customers No entity (including customers, mobile users, the service provider, or colluded entities) can link a sensing report to a specific mobile user or link a releasing task to a specific customer. It is even impossible for an attacker to identify whether two sensing reports are generated by the same mobile user or whether two tasks are issued by the same customer.
- Greedy user tracing Mobile users are prevented from submitting more than one report for the same task in a reporting period to obtain unfair rewards. The authority should trace the identity of greedy mobile users after receiving the transcripts of two different sensing reports.

4.2 PPFMC

We propose our PPFMC, which consists of four phases, setup, registration, task allocation and data reporting, based on the BBS+ signature [21] and the AFGH proxy re-encryption [20].

Setup The authority and the service provider bootstrap the MCS service. Let $(\mathbb{G}, \mathbb{G}_T)$ be two cyclic groups with a prime order p, where p is λ bits, and $\hat{e} : \mathbb{G} \times \mathbb{G} \rightarrow \mathbb{G}_T$ be a bilinear pairing. The authority generates random generators $g, g_0, g_1, g_2, g_3, h \in \mathbb{G}$, and calculates $G = \hat{e}(g, g)$ and $H = \hat{e}(h, h)$, respectively. The authority also randomly picks $\mathscr{G} \in \mathbb{G}_T$, and defines a secure hash function $\mathscr{H} : \{0, 1\}^* \rightarrow \mathbb{Z}_p^*$ and a pseudo-random function $\mathscr{F} : \mathbb{Z}_p \times \{0, 1\}^* \rightarrow \mathbb{Z}_p^*$. The public parameters are $(\mathbb{G}, \mathbb{G}_T, p, g, g_0, g_1, g_2, g_3, h, G, H, \mathscr{G}, \mathscr{H}, \mathscr{F})$. At last, the authority picks a random value $\alpha \in \mathbb{Z}_p^*$ as its secret key and generates the public key $T = g^\alpha$.

The service provider selects a random secret key $\beta \in \mathbb{Z}_p^*$ and calculates its public key $S = h^\beta$. It also utilizes a matrix $L_{m \times n}$ to denote the geographical region (i.e., longitude and latitude) that the MCS service can cover. Each entry in the matrix denotes a small grid in the sensing region. Given the longitude of Ontario is from 74.40°W to 95.15°W and the latitude is from 41.66°N to 57.00°N, we have a 208 × 154 matrix or a 2075 × 1534 matrix more precisely to represent it.

Each fog l randomly chooses $f_l \in \mathbb{Z}_p^*$ as the secret key and generates the corresponding public key as $F_l = h^{f_l}$. The secret key f_l is shared with the service provider through a secure channel.

Registration A customer or mobile user registers to the authority to acquire an anonymous credential. Each registrant (i.e., customer or mobile user) has a unique identity I, which is telephone number or mailing address in practise. The registrant chooses random values $s', a \in \mathbb{Z}_p^*$ to calculate $C = g_1^{s'} g_2^a$, $\widehat{A} = g^a$, and forwards $(I, C, \widehat{A})$ to the authority, along with the following zero-knowledge proof:

$$\mathscr{P}\mathscr{K}_1\{(s', a) : C = g_1^{s'} g_2^a \wedge \widehat{A} = g^a\}.$$

The authority checks the proof $\mathscr{PK}_1$ to guarantee $(C, \widehat{A})$ are properly conducted. The authority also randomly selects $s'', e \in \mathbb{Z}_p^*$ to compute $A = (g_0 C g_1^{s''} g_3^I)^{\frac{1}{\alpha+e}}$, $RK = \widehat{A}^{\frac{1}{\alpha}}$, and sends (A, s'', e, RK) to the registrant.

The registrant calculates $s = s' + s''$ and verifies

$$\hat{e}(A, Tg^e) \stackrel{?}{=} \hat{e}(g_0 g_1^s g_2^a g_3^I, g),$$

The registrant maintains $(A, e, s, a, I, \widehat{A}, RK)$ on his mobile device.

Task Allocation The time is divided into slots, each of which ranges from minutes to months depending on the specific requirements of the task. A customer with registered information $(A, e, s, a, I, \widehat{A}, RK)$ has a crowdsensing task to allocate and requests the sensing data from mobile users slot by slot. To protect the content privacy of the task, the customer selects four random values $k, r_1, r_2, r_3 \in \mathbb{Z}_p^*$, and computes $u = g^k, c_1 = S^{r_2}, c_2 = T^{r_1}$ and $c_3 = (task||u)G^{r_1}H^{r_2}$. Then, the customer computes a matrix $\widehat{L}_{m\times n}$ to denote the target sensing region $area$. The entry in $\widehat{L}_{m\times n}$ corresponding to each position in the sensing area is a random value in $\mathbb{Z}_p^*$; otherwise, the value for a location outside is set to be zero. To mask the sensing area in $\widehat{L}_{m\times n}$, the customer chooses $m \times n$ random numbers in $\mathbb{Z}_p^*$ to generate an invertible matrix $\widehat{M}_{m\times n}$ and computes $\widehat{N}_{n\times n} = \widehat{L}_{m\times n}^T \cdot \widehat{M}_{m\times n}$, where $\widehat{L}_{m\times n}^T$ is the transpose of the matrix $\widehat{L}_{m\times n}$. Note that all non-zero entries in $\widehat{L}_{m\times n}$ are distinct, unless an attacker can still learn the sensing region from $\widehat{N}_{n\times n}$. Lastly, the customer secretly stores k and forwards $(c_1, c_2, c_3, expires, \widehat{N}_{n\times n})$ to the service provider, along with the following zero-knowledge proof:

$$\mathscr{PK}_2\{(A, e, s, a, I) : \hat{e}(A, Tg^e)=\hat{e}(g_0 g_1^s g_2^a g_3^I, g)\}.$$

The service provider checks the validity of the proof $\mathscr{PK}_2$. If yes, it assigns a task number num, releases $(num, expires)$ and stores $(num, c_1, c_2, c_3, expires, \widehat{N}_{n\times n})$ in its database. The service provider finds the location of fogs and generates the matrix $\mathscr{F}_l$ for the fog l according to the location. The service provider computes $\widehat{N}_{n\times n} \times \mathscr{F}_l$ and $c_1' = c_1^{\frac{f_l}{\beta}}$, and allocates the task $(num, c_1', c_2, c_3, expires, \widehat{N}_{n\times n})$ to the fog l.

When a mobile user $U_{i\{i\in R\}}$ with $(A_i, e_i, s_i, a_i, I_i, \widehat{A}_i, RK_i)$ would like to engage in a crowdsensing task, U_i firstly randomly chooses $\nu \in \mathbb{Z}_p^*$ to compute $\mu = h^\nu$. Then, U_i generates a matrix $\widetilde{L}_{m\times n}$ based on its current location and the places it will visit. For each location where U_i will reach, the corresponding entry in $\widetilde{L}_{m\times n}$ is set as a random value chosen from $\mathbb{Z}_p^*$, and all other entries are zero. The non-zero entries in $\widetilde{L}_{m\times n}$ should be different. To protect the location, it also computes a random invertible matrix $\widetilde{M}_{m\times n}$ by choosing $m \times n$ random values from $\mathbb{Z}_p^*$, and computes $\widetilde{N}_{n\times n} = \widetilde{M}_{m\times n}^T \cdot \widetilde{L}_{m\times n}$. Finally, U_i keeps ν secretly and forwards $(\mu, \widetilde{N}_{n\times n})$ to the fog, along with the following zero-knowledge proof:

$$\mathscr{PK}_3\{(A_i, e_i, s_i, a_i, I_i) : \hat{e}(A_i, Tg^{e_i}) = \hat{e}(g_0 g_1^{s_i} g_2^{a_i} g_3^{I_i}, g)\}.$$

The fog returns failure if the verification of $\mathscr{PK}_3$ outputs invalid; otherwise, for each unexpired task, it utilizes $\widehat{N}_{n\times n}$ to compute $N_{n\times n} = \widetilde{N}_{n\times n} \cdot \widehat{N}_{n\times n}$ and verifies whether $N_{n\times n}$ is zero matrix or not. If $N_{n\times n}$ is non-zero matrix, U_i successfully matches ST. Then, the fog computes $c_4 = \hat{e}(\mu, c_1')^{\frac{1}{f_i}}$ and releases $(num, c_2, c_3, c_4, expires)$ for U_i. If there is no task matching U_i, the fog responds failure.

When obtaining $(num, c_2, c_3, c_4, expires)$, U_i decrypts (c_2, c_3, c_4) by using (ν, a_i), i.e., $task||u = c_3 c_4^{-\frac{1}{\nu}} \hat{e}(c_2, RK_i)^{-\frac{1}{a_i}}$. Then, U_i evaluates the task and determines to engage in or abandon this task based on the reward and cost. If U_i accepts the task ST, it performs the sensing work based on the details in $task$.

Data Reporting U_i senses the data m_i and periodically reports a sensing report to the customer. The reporting periods are specified by the customer. τ_j is denoted as the current slot. To protect m_i, U_i utilizes u to encrypt m_i as $D_i = u^{\hat{r}_i}$, $D_i' = m_i G^{\hat{r}_i}$, where $\hat{r}_i$ is randomly chosen from $\mathbb{Z}_p^*$. Then, U_i calculates $X_i = \mathscr{H}(num||m_i||\tau_j)$, $v_i = \mathscr{F}_{a_i}(num||I||\tau_j)$, $Y_i = H^{v_i}$ and $Z_i = \hat{e}(g, \widehat{A}_i)\mathscr{G}^{X_i v_i}$. Lastly, U_i forwards the report $(num, D_i, D_i', X_i, Y_i, Z_i, \tau_j)$ to the service provider.

The service provider verifies whether there is another report $(num, \widetilde{D}_i, \widetilde{D}_i', \widetilde{X}_i, Y_i, \widetilde{Z}_i, \tau_j)$ that has the same Y_i but different $\widetilde{X}_i$ with the new received report $(num, D_i, D_i', X_i, Y_i, Z_i, \tau_j)$. If yes, the service provider calculates $W = (\frac{\widetilde{Z}_i^{X_i}}{Z_i^{\widetilde{X}_i}})^{\frac{1}{X_i - \widetilde{X}_i}}$, and forwards it to the authority. Then, the authority finds the mobile user's identity I_i by using $\widehat{A}_i$ to verify whether $W = \hat{e}(g, \widehat{A}_i)$ or not, until it finds a match. Therefore, the identity of the greedy mobile user U_i^* is recovered by the authority if U_i^* submits two different sensing reports in a single reporting slot.

When retrieving the reports, the customer decrypts them with k as $m_i = D_i' \hat{e}(g, D_i)^{\frac{1}{k}}$, and distributes the rewards based on the contributions of mobile users.

4.3 Security Discussion

We explain the achieved security goals described in Sect. 4.1.3, including location privacy preservation, confidentiality of tasks and reports, anonymity of both mobile users and customers, and greedy user tracing.

- *Location Privacy Preservation* The location of the sensing region is represented as a matrix $\widehat{L}_{m\times n}$, which is randomized by a random matrix $\widehat{M}_{m\times n}$ to generate $\widehat{N}_{n\times n}$. The location of the mobile user is transformed to be $\widetilde{N}_{n\times n}$. Receiving these two matrices, the fog cannot learn any information about the location of the mobile user and the sensing area of the customer from $\widehat{N}_{n\times n}$ and $\widetilde{N}_{n\times n}$, respectively. The fog computes $N_{n\times n} = \widehat{N}_{n\times n} \cdot \widetilde{N}_{n\times n}$. If there is no overlapping in the crowdsensing area and location of users, $N_{n\times n}$ must be a zero matrix. If just one overlapping grid exists (where the corresponding entry is $\widehat{L}_{ij}$ in $\widehat{L}_{m\times n}$ and is $\widetilde{L}_{ij}$ in $\widetilde{L}_{m\times n}$, respectively), the entries in j-row of $\widehat{N}_{n\times n}$ are nonzero, as well as the entries in j-column of $\widetilde{N}_{n\times n}$. Then, the fog knows that there are some overlapping locations in the j-column of the sensing area. But, it is unable to distinguish which location is overlapped from m locations. Furthermore, $\widehat{N}_{n\times n} \cdot \widetilde{N}_{n\times n}$ and $\widetilde{N}_{n\times n} \cdot \widehat{N}_{n\times n}$ cannot provide more information to the fog. The results are the same if the overlapping grids are more than one. Therefore, the sensing area and the location of mobile user are not exposed to the fog and other entities.

- *Tasks Confidentiality* Having the location information matching the sensing area, the mobile users are able to decrypt the ciphertext of the task. In PPFMC, the adversaries may be the service provider, fogs, unmatched mobile users and external attackers. To resist these adversaries, the protection of the task consists of two stages. In the first stage, the task is encrypted by the customer under the public keys of the authority, the service provider and the fog; in the second stage, the fog partially decrypts the ciphertext using its secret key and then re-encrypts the result for the matched mobile users. Specifically, the security of the first-stage ciphertext can be reduced to the $q-$DBDHI assumption [20] and the security of the second-stage ciphertext depends on the AFGH proxy re-encryption scheme [20].
- *Reports Confidentiality* To guarantee the confidentiality of reports, each mobile user employs the AFGH proxy re-encryption scheme [20] to encrypt m_i under the temporary public key $u = g^k$, which is distributed to the mobile users along with the task. The decryption key k is secretly kept by the customer. Therefore, the confidentiality of m_i directly depends on the sematic security of the AFGH proxy re-encryption scheme, which can be reduced to the simplified $q-$DBDHI assumption [20].
- *Anonymity* The anonymity of the mobile user is defined via the game, in which the adversary cannot distinguish an honest mobile user out of two under the extreme condition that all other interactions are specified by the adversary. We show that the mobile user's identity is preserved properly, unless the DDH assumption [22] does not hold.
- *Greedy User Tracing* If there exist two sensing reports uploaded by the same mobile user in a time slot, the service provider can calculate a correct W and the authority can trace the identity of the greedy mobile user by checking the equation $W = \hat{e}(g, \widehat{A}_i)$.

4.4 Performance Evaluation

We evaluate the performance of our PPFMC in terms of computational and communication overhead.

4.4.1 Computational Overhead

We demonstrate the computational overhead of the PPFMC by counting the number of the basic cryptographic operations, such as point multiplication, point addition, bilinear map and exponentiation in $\mathbb{G}_T$. Other operations, e.g., multiplication in $\mathbb{G}_T$, addition, multiplication and inverse operations in $\mathbb{Z}_p^*$, can be negligible in computational time. We utilize the pre-processing mechanism to reduce the computational burden for each entity. Specifically, the authority pre-computes the bilinear maps $E_0 = \hat{e}(g_0, g)$, $E_1 = \hat{e}(g_1, g)$, $E_2 = \hat{e}(g_2, g)$, $E_3 = \hat{e}(g_3, g)$, $E_4 = \hat{e}(g_2, S)$ in the setup phase and $\{\hat{e}(g, \widehat{A}_i)\}_{i=0}^N$ in registration phase, where N is the number of registrants in the system. The mobile user U_i can also pre-compute $\hat{e}(g, \widehat{A}_i)$ in the registration phase. Table 4.1 shows the number of the operations required in each phase of PPFMC, respectively.

We also conduct an experiment to show the efficiency of PPFMC. The operations of authority and service provider are performed on a notebook with Intel Core i5-4200U CPU (the clock rate is 2.29 GHz) and 4.00GB memory. The operations of customers and mobile users are run on HUAWEI MT2-L01 smartphone with Kirin 910 CPU and 1250 M memory. The operation system is Android 4.2.2 and the toolset is Android NDK r8d. We use MIRACL library 5.6.1 to implement number-theoretic based methods of cryptography. The Weil pairing is utilized to realize the bilinear pairing operation. The parameter p is approximately 160 bits and the elliptic curve is defined as $y = x^3 + 1$ over $\mathbb{F}_q$, where q is 512 bits. The rough operation time of each entity in every phase of PPFMC is shown in Table 4.1.

4.4.2 Communication Overhead

We demonstrate the communication burden of PPFMC. The public parameters are the same as those in the experiment, that is, $|p|$=160 bits and $|q|$=512 bits. In the registration phase, a registrant (i.e., either customer or mobile user) sends a registering request $(I, C, \widehat{A}, \mathscr{PK}_1)$ to the authority. This registering request is of $|I| + 1344$ bits, where $|I|$ denotes the binary length of the identity. The authority returns (A, s'', e, RK) to the registrant, whose binary length is 1344 bits. In task allocation, the customer uploads $(c_1, c_2, c_3, expires, \widehat{N}_{n \times n}, \mathscr{PK}_2)$ and the mobile user sends $(\mu, \widetilde{N}_{n \times n}, \mathscr{PK}_3)$ to the fog, which are of $4512 + 160n^2 + |expires|$ bits and $2976 + 160n^2$ bits, respectively. Here, $|expires|$ denotes the binary length of $expires$. The service provider responds $(num, c_2, c_3, c_4, expires)$, which is $2560 + |num| + |expires|$ bits, to a matched mobile user or false, 1 bit, to an unmatched one, where $|num|$ denotes the binary length of num. After the mobile user obtains the sensing data, it sends the sensing report $(num, D_i, D_i', X_i, Y_i, Z_i, \tau_j)$ to the service provider. This sensing report is of $3768 + |num| + |\tau_j|$ bits, where $|\tau_j|$ denotes the binary length of τ_j. The service provider sends a 1024-bit W to the authority in case that a mobile user double-submits data, and forwards all the sensing reports to the customer.

Table 4.1 Computational Overhead of PPFMC

	Registration		Task allocation				Data reporting		
Phase	Authority	User	Customer	Provider	Fog	User	Customer	Provider	User
Point multiplication	9	10	11	12	0	9	0	0	1
Point addition	6	6	5	8	0	5	0	0	0
Bilinear map	0	2	1	5	1	2	1	0	0
Exponentiation in $\mathbb{G}_T$	0	0	8	13	1	8	1	3	3
Running time (ms)	45.524	151.304	98.701	120.715	15.343	147.685	56.405	1.283	3.661

4.5 Summary

In this chapter, we have introduced a privacy-preserving mobile crowdsensing framework that achieves location-based task allocation and preserves the identities of both customer and mobile users, simultaneously. In specific, we not only achieve identity privacy, location privacy and data privacy for both customers and mobile users, but also enable the service provider to recruit mobile users according to the points of interest of customers and the location of mobile users, Besides, we have discussed the security feature of conditional privacy preservation and analyzed the computational and communication overhead to demonstrate the efficiency of PPFMC.

References

1. J. Liu, H. Shen, H. S. Narman, W. Chung, and Z. Lin, "A survey of mobile crowdsensing techniques: A critical component for the internet of things," *ACM Transactions on Cyber-Physical Systems*, vol. 2, no. 3, p. 18, 2018.
2. A. Kamilaris and A. Pitsillides, "Mobile phone computing and the internet of things: A survey," *IEEE Internet of Things Journal*, vol. 3, no. 6, pp. 885–898, 2016.
3. R. K. Ganti, F. Ye, and H. Lei, "Mobile crowdsensing: current state and future challenges," *IEEE Communications Magazine*, vol. 49, no. 11, pp. 32–39, 2011.
4. Q. Xu, Z. Su, B. Han, D. Fang, Z. Xu, and X. Gan, "Analytical model with a novel selfishness division of mobile nodes to participate cooperation," *Peer-to-Peer Networking and Applications*, vol. 9, no. 4, pp. 712–720, 2016.
5. K. Yang, K. Zhang, J. Ren, and X. Shen, "Security and privacy in mobile crowdsourcing networks: challenges and opportunities," *IEEE communications magazine*, vol. 53, no. 8, pp. 75–81, 2015.
6. I. Krontiris, M. Langheinrich, and K. Shilton, "Trust and privacy in mobile experience sharing: future challenges and avenues for research," *IEEE Communications Magazine*, vol. 52, no. 8, pp. 50–55, 2014.
7. J. Ni, K. Zhang, X. Lin, Q. Xia, and X. Shen, "Privacy-preserving mobile crowdsensing for located-based applications," in *Proc. of ICC*, 2017, pp. 1–6.
8. J. Ni, "Security and privacy preservation in mobile crowdsensing," 2018.
9. S. Gisdakis, T. Giannetsos, and P. Papadimitratos, "Security, privacy, and incentive provision for mobile crowd sensing systems," *IEEE Internet of Things Journal*, vol. 3, no. 5, pp. 839–853, 2016.
10. A. Alamer, J. Ni, X. Lin, and X. Shen, "Location privacy-aware task recommendation for spatial crowdsourcing," in *Proc. of WCSP*, 2017, pp. 1–6.
11. C. Cornelius, A. Kapadia, D. Kotz, D. Peebles, M. Shin, and N. Triandopoulos, "Anonysense: privacy-aware people-centric sensing," in *Proc. of Mobisys*, 2008, pp. 211–224.
12. P. Gilbert, L. P. Cox, J. Jung, and D. Wetherall, "Toward trustworthy mobile sensing," in *Proc. of HotMobile*, 2010, pp. 31–36.
13. F. Qiu, F. Wu, and G. Chen, "Privacy and quality preserving multimedia data aggregation for participatory sensing systems," *IEEE Transactions on Mobile Computing*, vol. 14, no. 6, pp. 1287–1300, 2015.
14. K. Wang, X. Qi, L. Shu, D.-j. Deng, and J. J. Rodrigues, "Toward trustworthy crowdsourcing in the social internet of things," *IEEE Wireless Communications*, vol. 23, no. 5, pp. 30–36, 2016.

15. J. Ren, Y. Zhang, K. Zhang, and X. Shen, "Sacrm: Social aware crowdsourcing with reputation management in mobile sensing," *Computer Communications*, vol. 65, pp. 55–65, 2015.
16. J. Ni, X. Lin, K. Zhang, and Y. Yu, "Secure and deduplicated spatial crowdsourcing: A fog-based approach," in *Proc. of Globecom*, 2016, pp. 1–6.
17. L. Pournajaf, L. Xiong, V. Sunderam, and S. Goryczka, "Spatial task assignment for crowd sensing with cloaked locations," in *Proc. of MDM*, vol. 1, 2014, pp. 73–82.
18. J. Ni, K. Zhang, X. Lin, and X. Shen, "Securing fog computing for internet of things applications: Challenges and solutions," *IEEE Communications Surveys & Tutorials*, vol. 20, no. 1, pp. 601–628, 2017.
19. J. Ni, K. Zhang, Q. Xia, X. Lin, and X. Shen, "Enabling strong privacy preservation and accurate task allocation for mobile crowdsensing," *arXiv preprint arXiv:1806.04057*, 2018.
20. G. Ateniese, K. Fu, M. Green, and S. Hohenberger, "Improved proxy re-encryption schemes with applications to secure distributed storage," *ACM Transactions on Information and System Security (TISSEC)*, vol. 9, no. 1, pp. 1–30, 2006.
21. M. H. Au, W. Susilo, and Y. Mu, "Constant-size dynamic k-taa," in *Proc. of SCN*, 2006, pp. 111–125.
22. D. Cash, E. Kiltz, and V. Shoup, "The twin diffie-hellman problem and applications," in *Proc. of Eurocrypt*, 2008, pp. 127–145.

Chapter 5
Data Privacy Protection in Smart Grid

Smart grid enhances the power grid with information and communication technologies, such as control systems, network communication, and computation facilities, to enable two-way exchange of electricity and information between operation centers and smart meters, while making the grid more reliable, efficient, secure and greener [1]. In smart grid, operation centers are allowed to collect and analyze real-time power consumption and local energy generation for distribution management, outage identification, state estimation and dynamic billing. The operation centers share electricity consumption to power plants, thereby help power plants to adjust energy production and reduce the demand to fire up costly and secondary power plans [2]. Not only could the customers access real-time usage data and electricity prices, but also decrease their energy consumption by shifting the uninterrupted activities from peak time to non-peak time.

Even though the collected power consumption data help to promote the balance between supply and demand, it ends up with serious privacy issues toward customers, as it is possible to infer the customers' daily activities, habits and other privacy witnessable references from electricity usage data [3]. To be specific, a relatively low and static daily consumption of a household may indicate that no resident is at home [4]; a conspicuous drop of power consumption at midnight may indicate the family goes to sleep [5]. The prediction of behavior patterns is a serious privacy concern in smart grid defined by Electronic Privacy Information Center. To preserve customers' behavior patterns, IEC 62351 [6] recommends TLS encryption [7] to resist eavesdropping attack, including AES CBC, AES GCM or 3DES EDE CBC. Ontario Information Technology Standards employ IPsec or TLS to achieve authentication, integrity verification, privacy protection, and replay protection for advanced metering.

However, classical data encryption increases data size of consumption reports and brings heavy communication burden. To resolve these issues, privacy-preserving data aggregation schemes [8–10] have become increasingly popular to compress the consumption reports at local collectors (on behalf of fogs) and

X. Lin et al., *Privacy-Enhancing Fog Computing and Its Applications*,
SpringerBriefs in Electrical and Computer Engineering,
https://doi.org/10.1007/978-3-030-02113-9_5

forward the result in a compact form to the operation centers. These schemes can provide end-to-end confidentiality of meter readings, but cannot offer the integrity of consumption reports, indicating that they cannot provide sufficient integrity protection on consumption reports against misbehaving fogs [11]. The consumption reports are transmitted through public networks, e.g., Cellular network and the Internet, with the storage and forwarding of fogs, according to Toronto Hydro. Unfortunately, the fogs are vulnerable to be hacked. Malicious attackers may inject false data into the aggregated reports or corrupt the meter readings without being detected, and thereby affect state estimation, break power dispatch and control electricity prices. The electricity outage in Ukraine on Dec. 23, 2015 caused by a devastating cyberattack on a power station warns us that any vulnerability in advanced metering infrastructure may be exploited by hackers to create a blackout. Misbehaving fogs have not been received enough attentions lately. A handful of schemes [12, 13] aimed to reduce the dependence on a single fog or collector, but result in heavy communication burden to distribute the reliability to multiple fogs by means of secret sharing. Moreover, once the consumption reports of multiple customers are combined, it becomes a challenging task to support dynamic billing. In summary, it is of importance to achieve efficient smart metering simultaneously supporting data aggregation and dynamic billing with high security protection.

In this chapter, we design a Privacy-Preserving Smart Metering scheme (P^{2}SM) to realize end-to-end security, data aggregation and dynamic billing, simultaneously [14]. Considering a realistic case that the fogs at public areas may be controlled by adversaries, we propose a new security model between traditional semi-honest model and malicious model, which formally define the misbehavior of collectors. We realize identity authentication, data confidentiality and integrity of consumption reports against misbehaving fogs for smart metering based on several underlying techniques, including proxy re-encryption [15], Chameleon hash function [16] and homomorphic authenticators [17]. Besides, we upgrade the fogs with computing and storage resources, so that they are capable of temporarily maintaining individual consumption reports for dynamic billing. Specifically, our contributions are summarized as four-fold:

- Inspired by the fact that collectors or fogs in public areas may be hacked, we define a stronger security model to formalize their misbehavior in reality. Misbehaving fogs are not only curious about the behavior patterns of customers but will also insert false data into normal meter readings using pollution attacks to corrupt state estimation of operation centers.
- To prevent pollution attacks against fogs, we propose P^{2}SM by leveraging proxy re-encryption [15] and homomorphic authenticators [17]. The privacy-preserving data aggregation is achieved to prevent privacy leakage and reduce communication burden. P^{2}SM does not allow the collectors to generate their signatures by themselves, but aggregate the smart meters' signatures to guarantee the integrity of the aggregated consumption reports. As a result, a misbehaving fog cannot inject false data into the consumption reports or invade the privacy of customers.

- Once smart meters' signatures are aggregated, message authentication cannot be supported. We introduce an identity authentication mechanism based on the Chameleon hash function [16] for smart metering. With the desirable homomorphic property, authentication messages from multiple customers can be aggregated as well to further enhance communication efficiency.
- To enable dynamic billing, P^2SM allows fogs to utilize the individual consumption reports to generate verifiable daily bills for customers. Specifically, the collectors aggregate the consumption reports of each customer with the electricity prices to generate daily bills, and submit the result to the operation center. The operation center transforms the encrypted bills to be readable for customers. Moreover, the customers are able to check the correctness of their daily bills, and thereby discover the corruption of misbehaving fogs and greedy utilities.

5.1 Problem Statement

We formalize the system model, present security threats and identify design goals.

5.1.1 System Model

The system model is depicted in Fig. 5.1. Utilities have a sufficient supply of electricity from plants and offer power retailing service to customers. To achieve real-time power dispatch, operation centers collect and analyze real-time power consumption of customers and monitor power consumption by varying electricity prices. A temper-proof smart meter is installed in each customer's house to measure real-time electricity usage and submit the readings to operation centers every ρ minutes, $\rho = 15$ or 60 generally. A local fog, which is a wireless access point or base station, is deployed to connect the operation center and smart meters in a home area network. In each time slot, smart meters deliver meter readings to the fog through WiFi or ZigBee. After receiving the consumption reports, the fog transiently stores and aggregate them into a compact report, and delivers it to the operation center several times a day through a wired network, e.g., the Internet. The operation center monitors electricity distribution and determines dynamic electricity prices based on the total consumption of electricity. The electricity price is returned to the fog per day, and the fog generates the daily bills of customers based on the maintained real-time power consumption and sends the electricity bills to the utility. Finally, the customers view their electricity bills via the Internet and regulate their daily activities to decrease electricity costs.

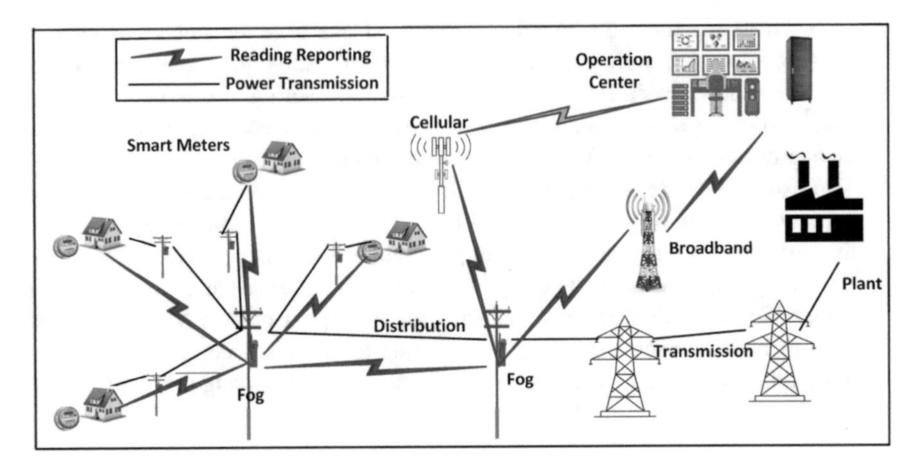

Fig. 5.1 System model for smart metering

5.1.2 Security Model

As the intermediates, local fogs are built in the public areas, and may be hacked by malicious hackers. The hackers have various misbehavior to invade customers' privacy, inject false data, corrupt state estimation and control electricity prices through these misbehaving collectors. To be as close to reality as possible, we define a new security model with a misbehaving adversary that has rational attack behaviors. The misbehaving fog may be neither completely malicious, to block the power usage data transmission, which can be quickly detected by the operation center, nor just honest-but-curious, to be curious about customers' living patterns. The compromised fog is more powerful than the honest-but-curious adversary, and more rational than the malicious adversary. On one hand, to prevent its misbehavior from being identified, a misbehaving fog will follow the communication protocols and pretend to be honest; while, it uses all sorts of methods to achieve its ulterior motives. In specific, a misbehaving fog utilizes the following attacks to invade customers' privacy and corrupt state estimation in smart grid:

- A misbehaving fog extracts the customers' privacy via eavesdropping.
- A misbehaving fog injects false data into power usage in home area network to corrupt state estimation or control electricity prices.
- A misbehaving fog may forge the smart meters' individual reports or aggregated reports to corrupt state estimation of the operation center.
- A misbehaving fog may forge the daily electricity bills to cheat the operation center, utilities and customers.

The eavesdropping and forgery attacks have been discussed in existing literatures [18, 19]. The pollution attack is a new type of attack. We define the following game between the misbehaving adversary and the advanced metering infrastructure to formally define this attack:

1. The advanced metering infrastructure setups the whole system to collect the power consumption of customers in a home area network.
2. The adversary interacts with the system and query the individual consumption reports, providing, for each query, a smart meter and its reading. The system generates the individual report for each query and returns the report to the adversary.
3. Finally, the adversary outputs an aggregated report different from the aggregation of queried individual consumption reports.

If the adversary generates a valid aggregated report which is not equal to the aggregation of queried individual reports with non-negligible probability, we say that the adversary wins the game. The smart metering scheme resists pollution attacks if the probability that the fog wins the above game is negligible.

The smart meters are physically protected to prevent customers from stealing electricity. The malfunction of smart meters would be discovered and replaced by utilities in time. Besides, the customers are honest to purchase the electricity from utilities and pay their daily electricity bills. The operation center, fully controlled by the government, is honest as well to perform power transmission and balance the power demand and response. The damage of the operation center directly impacts national security and social stability, and thus powerful security policies are deployed to protect the operation center. The utilities provide truthful power retailing services to their customers, while they are interested in customers' privacy and greedy on their benefits, increasing their income by modifying customers' bills.

5.1.3 Design Goals

To achieve privacy-preserving smart metering to resist the above security threats, P^2SM should have the following objectives:

- *Authentication* Individual consumption reports should be from legal customers. The attacker cannot deliver a forged consumption report acceptable for the operation center.
- *Privacy Preservation* No attacker can learn the meter readings and thereby invade the privacy of customers, even if attackers eavesdrop on communication channels or compromise the fogs. The curious utility cannot obtain the power consumption of their customers, except the daily bills.
- *Integrity Checking* Neither individual consumption reports nor aggregated reports can be modified by adversaries. Even the misbehaving fogs cannot corrupt the integrity of consumption reports by injecting false data into normal meter readings. Thus, the operation center obtains correct power usage data.

- *Dynamic Billing* The daily bills are generated from correct individual consumption reports and fluctuant electricity prices. The customers read their electricity bills and check their correctness.

5.2 P^{2}SM

We describe an overview of P^{2}SM to briefly show the work flow and information flow, and then give a detailed description of P^{2}SM.

5.2.1 Overview of P^{2}SM

The reason that the current privacy-preserving data aggregation schemes are vulnerable to the pollution attack is that the fog generates the signature on the aggregated consumption report to ensure report integrity. If the fog becomes dishonest, it can arbitrarily insert forged data into the aggregated report without being detected by the operation center. To resolve the pollution attack, we extend the homomorphic authenticators [17] to be pairing-based cryptosystem for supporting the aggregation of the signatures on various measurements generated by multiple smart meters, which is quite difficult if no parameter is pre-shared among smart meters [20]. To address this challenge, we first enable the smart meters to sign the meter readings rather than their ciphertexts with their secret keys using the homomorphic authenticators [17], and then allow the fogs to re-sign the individual signatures to generate signatures under a common key selected by the operation center based on bilinear pairing. In doing so, the re-signed signatures can be aggregated to prevent the misbehaving collector from corrupting the meter reading. Unfortunately, if the individual signatures are re-signed and aggregated, they cannot provide the authentication functionality for smart meters. To fix this drawback, we introduce a novel identity authentication scheme based on the Chameleon hash function [16] with batch verification, which results in a reduction of computational and communication overhead.

Further, the operation center cannot generate daily bills, after the individual consumption reports are compressed. To solve this issue, we introduce the idea of upgrading the capability of fogs with storage spaces. Hence, these individual reports transiently maintained on the fogs are used to compute the daily bills with the fluctuant electricity prices. To delegate the decryption of daily bills, the proxy re-encryption [15] is leveraged to enable the operation center to re-encrypt the bills generated by the fogs on behalf of a proxy. Due to the homomorphism [15], the ciphertexts of meter readings are able to be aggregated for improving the communication efficiency. In addition, to prevent the misbehaving fog from generating incorrect or fraudulent bills, the individual signatures of meter measurements are aggregated with the prices to generate verifiable tags on the daily bills. In this way, the customers can verify whether the daily bills are correctly conducted, and the corrupted ones can be identified.

P[2]SM is composed of six phases, System Initialization, Customer Registration, Report Generation, Report Aggregation, Report Reading and Dynamic Billing.

- System Initialization The operation center bootstraps the whole system and generates the public parameters $Params$ and its secret-public key pair (k, K).
- Customer Registration The customer C_i with a smart meter SM_i on the house registers at the operation center using the registration message $(SM_i, y_i, \mathscr{H}_i, z_{i1}, z_{i2})$, in which $\mathscr{H}_i$ is the commitment and (z_{i1}, z_{i2}) is the ciphertext of a random key k. The operation center returns (SM_i, RK_i) to the fog, where RK_i is the re-sign key utilized to transform C_i's signature to a signature under the common key α selected by the operation center for the smart meters in the home area network.
- Report Generation SM_i reads the measurement m_{it} at a time slot t and generates a consumption report $P_{it} = \mathscr{U}||SM_i||a'_{it}||c_{it}||e_{it}||\sigma_{it}||t$, in which a'_{it} is the authentication message, (c_{it}, e_{it}) is the ciphertext of m_{it} and σ_{it} is the signature on m_{it}. SM_i sends P_{it} to the fog.
- Report Aggregation The fog aggregates the authentication messages, ciphertexts and signatures in the individual consumption reports during each forwarding period Q to generate an aggregated report $P = \mathscr{C}||a||c||e||\sigma||Q$ using the re-sign keys RK_i of all smart meters in its home area network, and forwards P to the operation center.
- Report Reading The operation center verify the validity of the aggregated authentication message a, decrypts the aggregated ciphertext (c, e) and verifies the aggregated signature σ. Finally, the operation center obtains the total power consumption m for state estimation and demand response.
- Dynamic Billing The operation center determines the fluctuant electricity prices $(p_1, \cdots, p_\varphi)$ during a day. The fog aggregates the ciphertexts of meter readings with the prices to generate the daily bill (c_i, e_i) for the customer, and aggregates the signatures with the prices to obtain a verifiable tag τ_i on the daily bill. To allow the customer to read the bill, the operation center re-encrypts the bill $B_i = y_i||SM_i||\mathscr{U}||c_i||e_i||\tau_i$ to generate $B'_i = y_i||SM_i||\mathscr{U}||l_i||e_i||\tau_i$ for the customer. Thus, the customer reads the daily bill and utilizes τ_i to verify the correctness of the daily bill.

5.2.2 *The Detailed P[2]SM*

We then describe the construction of P[2]SM in detail.

5.2.2.1 System Initialization

The operation center (OC) provides electricity distribution and demand response for customers $\mathbb{C} = \{C_1, \cdots, C_N\}$ in the residential area $\mathbb{RA}$. Suppose that $\mathbb{C}$ purchases electricity from a utility $\mathscr{U}$, (it is compatible that $\mathbb{C}$ buys the power offered by multiple utilities). OC bootstraps the advanced metering infrastructure on behalf. Specifically, OC first chooses the security parameter κ, which denotes the security level and κ is 160 or 256 usually. OC picks a large prime p, where $|p| = \kappa$. OC generates two cyclic groups $(\mathbb{G}, \mathbb{G}_T)$ with the same order p. g, g_0 are generators of $\mathbb{G}$, and $\hat{e} : \mathbb{G} \times \mathbb{G} \rightarrow \mathbb{G}_T$ is a bilinear pairing. $H : \{0,1\}^* \rightarrow \mathbb{G}$ and $H_1 : \{0,1\}^* \rightarrow \{0,1\}^\kappa$ are cryptographic hash functions and $F : \mathbb{G} \times \{0,1\}^* \rightarrow \mathbb{Z}_p$ is a pseudo-random function. (E_s, D_s) are the encryption and decryption algorithms of AES. Then, OC randomly picks $k \in \mathbb{Z}_p$ to generate $K = g^k$. Finally, OC releases the public parameters:

$$Params = \{p, \mathbb{G}, \mathbb{G}_T, \hat{e}, g, g_0, H, H_1, F, E_s, D_s, K\},$$

and keeps the secret key k in private.

5.2.2.2 Customer Registration

When a customer $C_i \in \mathbb{C}$'s house in the $\mathbb{RA}$ connects with a smart grid, OC installs a smart meter SM_i for C_i. In the registration, C_i first randomly picks $x_i \in \mathbb{Z}_p$ as the private key and generates the corresponding public key as $y_i = g^{x_i} \in \mathbb{G}$. The secret key x_i is plugged into SM_i or stored in a trusted platform module (TPM) integrated into SM_i and the public key certificate is publicly accessible. Then, SM_i randomly chooses $a_i, b_i \in \mathbb{Z}_p$ to compute a Chameleon hash value $\mathscr{H}_i = g^{a_i} y_i^{b_i}$. After that, SM_i picks two random values $k_i, r_i \in \mathbb{Z}_p$ to calculate $z_{i1} = g^{r_i}, r_i' = H_1(z_{i1}, K^{r_i})$ and $z_{i2} = E_s(r_i', k_i)$. Lastly, SM_i sends $(SM_i, y_i, \mathscr{H}_i, z_{i1}, z_{i2})$ to OC, and keeps (a_i, b_i, k_i) in the TPM, along with x_i.

Upon receiving $(SM_i, y_i, \mathscr{H}_i, z_{i1}, z_{i2})$, OC decrypts (z_{i1}, z_{i2}) to acquire the tag k_i as $r_i' = H_1(z_{i1}, z_{i1}^k)$ and $k_i = D_s(r_i', z_{i2})$. Then, OC randomly chooses $\alpha \in \mathbb{Z}_p$ as a unique identifier of $\mathbb{RA}$ to compute a re-sign key $RK_i = y_i^\alpha$, if C_i is the first customer in $\mathbb{RA}$; otherwise, U_i uses the existing α to compute RK_i. At last, OC sends (SM_i, RK_i) to the local fog in $\mathbb{RA}$ through a secure channel, and stores $(SM_i, y_i, \mathscr{H}_i)$ in its database and keeps $(T_i, RK_i, \alpha, g^\alpha)$ secretly.

5.2.2.3 Report Generation

To achieve real-time power dispatch, smart meters measure power consumption and deliver electricity consumption reports every ρ minutes, i.e., $\rho = 15$ or 60 ($\rho = 60$

for Toronto Hydro). Suppose that a smart meter SM_i measures the meter reading m_{it} at a time slot t. SM_i generates an individual consumption report as follows:

- Use $b'_{it} = F(H(SM_i||k_i), t)$ to generate the authentication message $a'_{it} = x_i \cdot (b_i - b'_{it}) + a_i \mod p$;
- Randomly choose $s_{it} \in \mathbb{Z}_p$ to compute the ciphertext of m_{it} as:

$$c_{it} = K^{s_{it}}, \quad e_{it} = \hat{e}(g_0^{m_{it}} g^{s_{it}}, g);$$

- Use x_i to obtain a signature as:

$$\sigma_{it} = (H(SM_i||\mathscr{U}||t) g_0^{a'_{it}} g^{m_{it}})^{\frac{1}{x_i}}; \tag{5.1}$$

- Send the individual consumption report $P_{it} = \mathscr{U}||SM_i||a'_{it}||c_{it}||e_{it}||\sigma_{it}||t$ to the fog in this area.

5.2.2.4 Report Aggregation

The fog transiently maintains the received individual consumption reports. It forwards the consumption reports to OC φ times per day ($\varphi = 5$ or 24). In each forwarding period Q, the fog aggregates the individual consumption reports received in Q from N smart meters in $\mathbb{RA}$ into an aggregated report P as follows:

$$c = \prod_{t \in Q} \prod_{i=1}^{N} c_{it}; \qquad e = \prod_{t \in Q} \prod_{i=1}^{N} e_{it}; \tag{5.2}$$

$$a = \sum_{t \in Q} \sum_{i=1}^{N} a'_{it} \mod p; \quad \sigma = \prod_{t \in Q} \prod_{i=1}^{N} \hat{e}(\sigma_{it}, RK_i). \tag{5.3}$$

The collector sets $P = \mathscr{C}||a||c||e||\sigma||Q$ and forwards P to OC, where $\mathscr{C}$ is the identifier of the fog.

5.2.2.5 Report Reading

After receiving $P = \mathscr{C}||a||c||e||\sigma||Q$, OC conducts the following steps to read the aggregated report P:

- Use each customer's unique tag k_i to compute $b^*_{it} = F(H(SM_i||k_i), t)$ and verify whether all reports are released by legitimate smart meters by checking Eq. (5.4):

$$\prod_{t\in Q}\prod_{i=1}^{N}\mathscr{H}_i \stackrel{?}{=} g^a \cdot \prod_{t\in Q}\prod_{i=1}^{N} y_i^{b_{it}^*}. \tag{5.4}$$

If Eq. (5.4) holds, continue to decrypt (c, e); otherwise, retrieve the individual consumption reports from the fog to discover the invalid reports.

- Decrypt the aggregated ciphertext (c, e) as $M = e\hat{e}(c, g)^{-\frac{1}{k}}$ and recover the discrete log of M base $\hat{e}(g_0, g)$ using Pollard's lambda method [21] to acquire $m = \sum_{t\in Q}\sum_{i=1}^{N} m_{it}$.
- Verify whether Eq. (5.5) is valid or not:

$$\sigma \stackrel{?}{=} \hat{e}(\prod_{t\in Q}\prod_{i=1}^{N} H(SM_i||\mathscr{U}||t)g_0^a g^m, g^\alpha). \tag{5.5}$$

If yes, accept m, which is the total power consumption in $\mathbb{RA}$ in the period Q; otherwise, reject m and retrieve the individual consumption reports from the fog to find the corrupted reports utilizing a recursive divide-and-conquer approach.

5.2.2.6 Dynamic Billing

According to the power consumption of customers in $\mathbb{RA}$, OC determines the electricity price in every forwarding period during a day, that is, $(p_1, \cdots, p_\varphi)$, where p_j denotes the electricity price in the jth forwarding period Q_j, and sends $(p_1, \cdots, p_\varphi)$ to the collector. The collector aggregates the individual consumption reports of a customer with the electricity prices to generate a daily bill for the customer. Specifically, for a customer C_i, the fog computes

$$c_i = \prod_{j=1}^{\varphi}\prod_{t\in Q_j} c_{it}^{p_j},\ e_i = \prod_{j=1}^{\varphi}\prod_{t\in Q_j} e_{it}^{p_j},\ \tau_i = \prod_{j=1}^{\varphi}\prod_{t\in Q_j} \sigma_i^{p_j}, \tag{5.6}$$

where $t \in Q_j$ means that the time slot t is in the reporting period Q_j, and sends the bill $B_i = y_i||SM_i||\mathscr{U}||c_i||e_i||\tau_i$ to OC. Then, OC verifies the correctness of the bills in $\mathbb{RA}$ by verifying Eq. (5.7):

$$\hat{e}(g_0, g)^{\sum_{j=1}^{\varphi} m_j p_j} = \prod_{i=1}^{N} e_i\hat{e}(c_i, g)^{-\frac{1}{k}}, \tag{5.7}$$

where m_j is the total power consumption of customers in $\mathbb{RA}$ in the period Q_j. If it holds, OC further computes $l_i = \hat{e}(c_i, y_i)^{\frac{1}{k}}$ and delivers the bill $B_i' = y_i||SM_i||\mathscr{U}||l_i||e_i||\tau_i$ to $\mathscr{U}$. Besides, OC delegates $\mathscr{U}$ to conduct proxy re-encryption to transform the ciphertexts of OC to be decryptable for C_i on behalf of a proxy. Specifically, OC sends $USK_i = y_i^{\frac{1}{k}}$ to $\mathscr{U}$ to compute $l_i = \hat{e}(c_i, USK_i)$ for C_i. Finally, C_i decrypts (l_i, e_i) by computing $D_i = e_i l_i^{-\frac{1}{x_i}}$ and recovering the discrete log of D_i base $\hat{e}(g_0, g)$ using Pollard's lambda method [21] to obtain $d_i = \sum_{j=1}^{\varphi} \sum_{t \in Q_j} m_{it} p_j$. To verify the correctness of d_i, C_i checks Eq. (5.8):

$$\hat{e}(\tau_i, y_i) \stackrel{?}{=} \hat{e}(\prod_{j=1}^{\varphi} \prod_{t \in Q_j} H(SM_i||\mathscr{U}||t)^{p_j} g_0^{\sum_{j=1}^{\varphi} \sum_{t \in Q_j} a'_{it} p_j} g^{d_i}, g), \tag{5.8}$$

where $a'_{it} = x_i(b_i - F(H(SM_i||k_i), t) + a_i) \mod p$. If Eq. (5.8) holds, C_i accepts the bill d_i; otherwise, rejects it.

5.3 Security Analysis

We analyze the security properties of P[2]SM, including authentication, confidentiality and integrity.

- *Authentication* SM_i utilizes the ElGamal encryption to send k_i to OC. Since the ElGamal encryption is semantically secure against chosen plaintext attacks [22] based on Hash-Diffie-Hellman problem, only OC can recover k_i. Thus, k_i is shared between SM_i and OC. To realize efficient authentication, the Chameleon hash function is leveraged to realize the interactions between SM_i and OC. Firstly, $\mathscr{H}_i = g^{a_i} y_i^{b_i}$ is one-way, which means that $\mathscr{H}_i$ can be computed from (a_i, b_i), but no one can extract (a_i, b_i) from $\mathscr{H}_i$, if the Discrete Logarithm (DL) assumption [16] holds. Besides, the attacker cannot find a collision (a'_i, b'_i) of (a_i, b_i) to make $\mathscr{H}_i = g^{a'_i} h_i^{b'_i}$ hold without x_i in polynomial time with non-negligible probability, unless the DL problem is tractable. However, having x_i, SM_i is able to generate a_i from any given b_i. Thus, if the ElGamal encryption is semantically secure and the DL problem is intractable, it is difficult for an adversary to pretend to be a legal smart meter to generate consumption reports without being detected by OC.
- *Confidentiality* To prevent the leakage of customers' power consumption, we adopt to the proxy re-encryption [15] to encrypt meter readings. Since the proxy re-encryption is proved secure against chosen plaintext attacks, the confidentiality of m_{it} is satisfied, such that attackers cannot invade C_i's privacy, even attackers eavesdrop and capture the ciphertexts. When the fog obtains all individual consumption reports in $\mathbb{RA}$, it cannot recover the meter readings but aggregating the ciphertexts to reduce communication overhead. As for OC, it is able to recover the sum of the power consumed in $\mathbb{RA}$ with its secret key.

In dynamic billing phase, the fog aggregates the individual consumption reports with the electricity prices to generate the bills and forwards the result to OC or U to allow them to re-encrypt the bills to be decryptable for the customers. Since the proxy re-encryption is secure under the Computational Bilinear Inverse Diffie-Hellman (BIDH) assumption [15], the meter readings and the electricity bills are kept confidential.

- *Integrity* To resist pollution attacks, P^{2}SM ensures the integrity of consumption reports from smart meters to the operation center. The signatures of smart meters can ensure the integrity of individual consumption reports, and the aggregated signature can prevent data corruption during the transmission from the collector to the operation center. In this way, the integrity of reports relies on the unforgeability of both the individual signatures and the aggregated signature. We will prove the unforgeability of the individual signatures and the aggregated signature separately.

To ensure the integrity of the individual consumption report, SM_i generates a signature with its private key as $\sigma_{it} = (H(SM_i||\mathcal{U}||t)g_0^{a'_{it}}g^{m_{it}})^{\frac{1}{x_i}}$. The unforgeability of this signature is reduced to the Diffie-Hellman Inversion (DHI) assumption [23], that is, within non-negligible advantage, no probabilistic polynomial-time algorithm can solve DHI problem: given $h, h^s \in \mathbb{G}$, where $s \in \mathbb{Z}_p$, to compute $h^{\frac{1}{s}} \in \mathbb{G}$.

Theorem 1 *The signature in an individual consumption report is existentially unforgeable against adaptive chosen message attacks under the security model* [19], *provided that the DHI problem is intact with a non-negligible probability in probabilistic polynomial time.*

Proof Suppose that an adversary $\mathcal{A}$ breaks the existential unforgeability of the signature with a non-negligible probability, then we can construct an algorithm $\mathcal{B}$ to solve the DHI problem. Let h be a generator of $\mathbb{G}$. $\mathcal{B}$ is given $h, h^s \in \mathbb{G}$, where $s \in \mathbb{Z}_p$, its goal is to output $h^{\frac{1}{s}}$. $\mathcal{B}$ simulates a challenger and interacts with $\mathcal{A}$ in the following way.

- In the setup, $\mathcal{B}$ sets the public key v to $h^{\frac{s}{r}}$ and the parameters g to h^{sr_1}, g_0 to h^{sr_2}, where r, r_1, r_2 are randomly picked in $\mathbb{Z}_p$, and forwards them to $\mathcal{A}$.
- $\mathcal{B}$ programs a random oracle to respond to hash queries. To keep the consistency, it manages a list of tuples to keep the queries and corresponding results. When receiving queries $(SM_i, \mathcal{U}, t)$ from $\mathcal{A}$, $\mathcal{B}$ flips a bias coin $\theta_i \in \{0, 1\}$, such that $\Pr[\theta_i = 0] = 1/(q_s + 1)$, where q_s is the maximum of signing queries that $\mathcal{A}$ can make. If $\theta_i = 0$, $\mathcal{B}$ computes $w_i = h^{\beta_i}$; otherwise, $\theta_i = 1$ and $\mathcal{B}$ computes $w_i = h^{s\beta_i}$, where β_i is randomly selected from $\mathbb{Z}_p$. Finally, $\mathcal{B}$ inserts a tuple $(SM_i, \mathcal{U}, t, \theta_i, \beta_i, w_i)$ into the list, and returns w_i to $\mathcal{A}$.

- $\mathscr{B}$ also programs a signing oracle and manages a list of tuples to stores the queries and responses. When $\mathscr{A}$ queries $(SM_i, \mathscr{U}, t, a'_{it}, m_{it})$, $\mathscr{B}$ firstly checks the list in hash queries. If $(SM_i, \mathscr{U}, t)$ has not been queried, $\mathscr{B}$ generates the corresponding (θ_i, β_i, w_i) for $(SM_i, \mathscr{U}, t)$. If $\theta_i = 0$, $\mathscr{B}$ aborts and returns failure; If $\theta_i = 1$, $\mathscr{B}$ sets $\sigma_{it} = h^{\beta_i r + r_1 a'_{it} r + r_2 m_{it} r}$. Observe that σ_{it} is a valid signature on $(SM_i, \mathscr{U}, t, a'_{it}, m_{it})$ with the public key $h^{\frac{s}{r}}$. Finally, $\mathscr{B}$ responds σ_{it} and inserts $(SM_i, \mathscr{U}, t, \theta_i, \beta_i, a'_{it}, m_{it}, \sigma_{it})$ into the list.
- Eventually, $\mathscr{A}$ outputs a message-signature pair $(SM_i, \mathscr{U}, \hat{t}, \hat{a}'_i, \hat{m}_i, \hat{\sigma}_i)$, and no signature query has been made for $(SM_i, \mathscr{U}, \hat{t}, \hat{a}'_i, \hat{m}_i)$. If no tuple is in the hash list, $\mathscr{B}$ issues $(SM_i, \mathscr{U}, \hat{t})$ to hash query. $\mathscr{B}$ aborts and returns failure, if $\hat{\sigma}_i$ is invalid. Then, $\mathscr{B}$ finds the tuple on hash list. If $\hat{\theta}_i = 1$, $\mathscr{B}$ aborts and returns failure; otherwise, $\hat{\theta}_i = 0$ and therefore $H(SM_i||\mathscr{U}||\hat{t}) = h^{\hat{\beta}_i}$. Hence, $\hat{\sigma}_i = h^{\frac{\hat{\beta}_i r}{s}} h^{r_1 \hat{a}'_i r + r_2 \hat{m}_i r}$. Then, $\mathscr{B}$ outputs the required $h^{\frac{1}{s}} = (\hat{\sigma}_i h^{-(r_1 \hat{a}'_i r + r_2 \hat{m}_i r)})^{\frac{1}{\hat{\beta}_i r}}$

Therefore, if the DHI problem cannot be solved with a non-negligible probability in probabilistic polynomial time, no adversary can forge the signatures on individual reports.

The integrity of bills is reduced to the DHI assumption, since the signatures on bills are the aggregation of smart meter's signatures. If the single signature σ_{it} is unforgeable, it is hard to forge its aggregated signature τ_i.

Then, we demonstrate that it is difficult to forge a valid aggregated report-signature pair (σ, m) in probabilistic polynomial time under the assumption of Conference-Key Sharing (CONF) [24] in group $\mathbb{G}_T$, that is, there is no probabilistic polynomial-time algorithm that solves the CONF problem [24] within a non-negligible probability: given $g, g^s g^{sv} \in \mathbb{G}$, where $s, v \in \mathbb{Z}_p$, to compute $\hat{e}(g, g)^v \in \mathbb{G}_T$.

Theorem 2 *The probability of conducting a valid aggregated signature σ, which is not equal to the aggregation of smart meters' signatures, in probabilistic polynomial time is negligible, provided that the CONF problem is intact.*

Proof If a probabilistic polynomial-time adversary $\mathscr{A}$ can break the unforgeability of the aggregated signature within a non-negligible probability, we can construct an algorithm $\mathscr{B}$ to solve the CONF problem.

Let g be a generator of $\mathbb{G}$. $\mathscr{B}$ is given $g, D = g^s, D_1 = g^{sv} \in \mathbb{G}$, where $s, v \in \mathbb{Z}_p$, its goal is to output $D_2 = \hat{e}(g, g)^v$. $\mathscr{B}$ simulates a challenger, who can access the signing oracle $\mathscr{SO}$ that outputs the signatures on individual reports, and interacts with the adversary $\mathscr{A}$ as follows.

- $\mathscr{B}$ randomly selected $r_i \in \mathbb{Z}_p$ to set the public key h_i to D^{r_i} and the re-sign key RK_i to $D_1^{r_i}$, for $1 \leq i \leq N$. $\mathscr{B}$ also randomly chooses $\gamma \in \mathbb{Z}_p$ to generate the parameter g_0 to g^γ. Lastly, $\mathscr{B}$ sends $(\{h_i, RK_i\}_{1 \leq i \leq N}, g, g_0)$ to $\mathscr{A}$.

- $\mathscr{A}$ queries the signatures on individual reports under any public key in $\{h_i\}$ for $1 \leq i \leq N$. $\mathscr{B}$ issues a signing query to $\mathscr{SO}$ and receives σ_i, and returns σ_i to $\mathscr{A}$.
- Finally, $\mathscr{A}$ produces an aggregated signature $\hat{\sigma}$ on the compressed reports $(SM_i, \mathscr{U}, \hat{t}, \hat{a}, \hat{m})$ for $1 \leq i \leq N$ in a time period $t \in Q$. Suppose that $\hat{\sigma}$ is a valid signature on $\hat{m}$; otherwise, $\mathscr{B}$ reports failure and aborts. Thus, $\hat{\sigma}$ satisfies the verification equation, i.e.,

$$\hat{\sigma}=\hat{e}(\prod_{t\in Q}\prod_{i=1}^{N} H(SM_i||\mathscr{U}||t)g^{\hat{a}}g_0^{\hat{m}}, g^v).$$

Assume the expected signature, which would be obtained from the honest smart meters, be σ on the report $(SM_i, \mathscr{U}, t, a, m)$ for $1 \leq i \leq N$ in a time period $t \in Q$. σ also satisfies the verification equation, i.e.,

$$\sigma=\hat{e}(\prod_{t\in Q}\prod_{i=1}^{N} H(SM_i||\mathscr{U}||t)g^{a}g_0^{m}, g^v).$$

If $\hat{a} = a$ and $\hat{m} = m$, then $\hat{\sigma} = \sigma$. Define $\Delta a = \hat{a} - a$ and $\Delta m = \hat{m} - m$, then, either Δa or Δm is nonzero.

- If $\hat{\sigma} \neq \sigma$, we divide the verification equation for $\hat{\sigma}$ by the equation for σ and obtain

$$\hat{\sigma}/\sigma=\hat{e}(\prod_{t\in Q}\prod_{i=1}^{N} g^{\Delta a}g_0^{\Delta m}, g^v).$$

Since $g_0 = g^\gamma$, we have

$$\hat{\sigma}/\sigma=\hat{e}(g^{\sum_{t\in Q}\sum_{i=1}^{N}\Delta a+\gamma\Delta m}, g^v).$$

Rearranging the equation yields

$$D_2 = \hat{e}(g, g)^v = (\frac{\sigma}{\hat{\sigma}})^{\sum_{t\in Q}\sum_{i=1}^{N}\Delta a+\gamma\Delta m},$$

which is the solution to the CONF problem.

– Otherwise, we get $\hat{e}(g^{\sum_{t\in Q}\sum_{i=1}^{N}\Delta a+\gamma\Delta m}, g^{v}) = 1$ and

$$D_2 = \hat{e}(g, g)^{v} = 1^{\sum_{t\in Q}\sum_{i=1}^{N}\Delta a+\gamma\Delta m}.$$

So we can solve the CONF problem.

Therefore, if the CONF problem cannot be addressed with a non-negligible probability in probabilistic polynomial time, any adversary cannot corrupt the aggregated reports.

In summary, P^{2}SM achieves authentication, confidentiality and integrity of the individual consumption reports and aggregated reports, as well as the electricity bills. Therefore, the misbehaving fog cannot corrupt consumption reports or electricity bills without being identified.

5.4 Performance Evaluation

We evaluate the performance of P^{2}SM with respect to computational, communication and storage overhead.

5.4.1 Computational Cost

To demonstrate the computational efficiency of P^{2}SM, we count the number of time-consuming operations on elliptic curve groups, including scalar multiplication (SM), point addition (PA), hash to point (HP), bilinear pairing (BP) and multiplication in $\mathbb{G}_T$ (MU_T). When C_i registers on OC, it performs $3SM + HP + 2PA$ to obtain $(\mathscr{H}_i, z_{i1}, z_{i2})$, and OC executes $2SM+PA$ to acquire k_i and conducts RK_i. In each t, SM_i generates $(a'_{it}, c_{it}, t_{it}, \sigma_{it})$ to obtain P_{it}, in which SM_i executes $4SM + HP + 2PA + 2MU_T$. $\hat{e}(g_0, g)$ and $\hat{e}(g, g)$ are pre-computed to reduce the computational burden for SM_i. The fog runs $(2N|Q|-2)PA+N|Q|BP$ to generate P. Finally, OC performs $(2N|Q|+1)PA+(N|Q|+4)SM+N|Q|HP+MU_T+BP$ and discrete logarithm to acquire the sum of power consumption in $\mathbb{RA}$, if all the reports are valid. Otherwise, OC executes $N|Q|(2SM + PA)$ to discover the invalid authentication messages if (5.4) does not hold; or $N|Q|(2PA + 2SM + HP + 2BP)$ to find the invalid signatures if (5.5) does not hold. In dynamic billing phase, for C_i, the fog aggregates the power consumption by performing $72|Q_j|SM+(24|Q_j|-3)PA$. Then, OC executes $NSM+(2N-1)MU_T$ to check the correctness of electricity bills and executes $SM + BP$ operations to obtain l_i or

$\mathcal{U}$ helps OC to compute BP. Lastly, C_i executes MU_T and discrete logarithm to recover the bill d_i, and checks the correctness of d_i by executing $(24|j|+2)SM+2PA$ $+24|j|HP+2BP$.

Table 5.1 Comparison of time costs

Unit: ms, $N = 100$						
	Report generation	Report aggregation	Report reading	Dynamic billing		
Phase	Smart meter	Fog	Operation center	Operation center	Customer	Fog
P^{2}SM	24.7	2355.4	328.1	3564.1	54.9	2254.7
EPPA [8]	26.2	4906.3	97.7	Null	Null	Null
Fan14 [25]	7.3	6737.6	130.9	Null	Null	Null
Ohara14 [26]	96.2	25.4	174.3	26.5	183.4	Null

We conduct an experience on a notebook with Intel Core i5-4200U CPU @ 2.29 GHz and 4.00 GB memory. We use the MIRACL library to implement number-theoretic based methods of cryptography. The Weil pairing is utilized to realize the bilinear pairing and the elliptic curve is a base field size of 512 bits. p is 160 bits. We compare P^{2}SM with three existing schemes, EPPA [8] (based on Paillier encryption [27]), Fan14 [25] (based on BGN encryption [18]) and Ohara14 [26] (based on Lifted ElGamal encryption [22]). While P^{2}SM is designed from proxy re-encryption [15]. To keep the consistency, we utilize the same settings in the experience. The number of customers in $\mathbb{RA}$ is 100, the number of reporting slots in a period Q is 1 and the number of reporting period per day φ is 24. The execution time of each entity in each phase of four schemes are demonstrated in Table 5.1. Figure 5.2 exhibits the comparison results of four schemes. Although Fan14 [25] is most efficient in report generation, as it utilizes BGN to encrypt the meter readings and no commitment is required, it is the most inefficient one in report aggregation. EPPA [8] is fastest in report reading, since no discrete logarithm computation is required and less

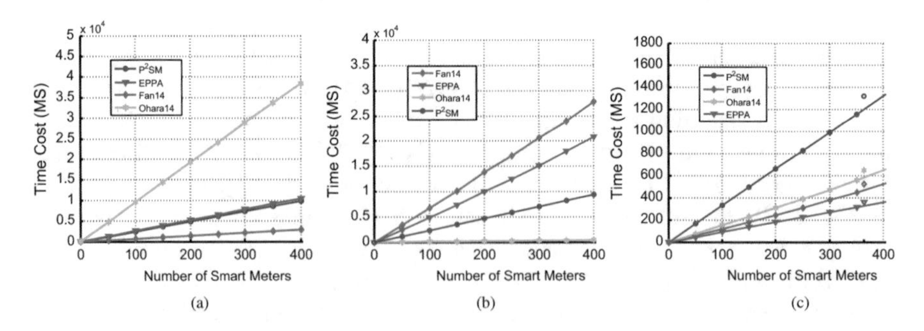

Fig. 5.2 Comparison on computational overhead. (**a**) Computational cost of smart meters. (**b**) Computational cost of the fog. (**c**) Computational cost of the operation center

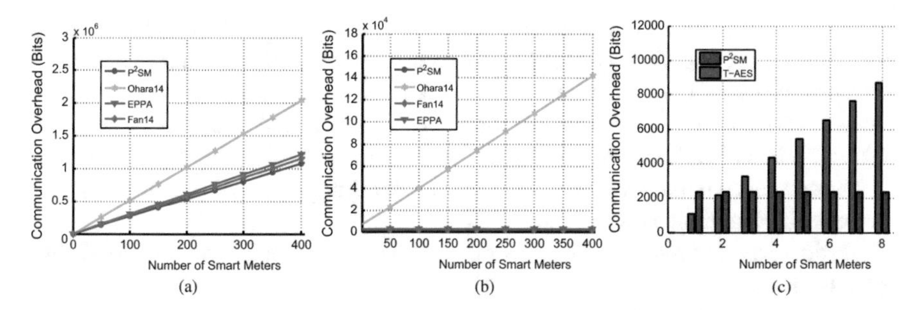

Fig. 5.3 Comparison on communication overhead. (**a**) Overhead between SM and fog. (**b**) Overhead between fog and OC. (**c**) Comparison of P^{2}SM and TLS protocol

bilinear pairings are conducted compared with P^{2}SM, while it is time-consuming in report aggregation. Future, Fan14 [25] and EPPA [8] cannot support dynamic billing. Ohara14 [26] has a good performance on the computational overhead, but it is vulnerable to the pollution attacks, as well as Fan14 [25] and EPPA [8]. P^{2}SM is not the most efficient in four schemes, even the least efficient one in report reading, because it uses the pairing-based cryptosystem to offer a higher security guarantee compared with Fan14 [25], EPPA [8] and Ohara14 [26]. Besides, the bottleneck of the computational efficiency is the smart meters, and our scheme is quite efficient on report generation for smart meters. P^{2}SM is still efficient since the time cost on report reading is only 328ms, while the operation center is always powerful. The fog's computational burden can be reduced by decreasing the number of smart meters in its coverage area and building the hierarchical report collection to improve the efficiency of smart metering.

5.4.2 Communication and Storage Overhead

The communications of P^{2}SM composes of SM-to-fog communication and fog-to-OC communication. In SM-to-fog, SM_i sends 2688-bit P_{it} at a time slot t to the fog, where $\mathscr{U}$, SM_i and t are assumed to be 160 bits, respectively. In fog-to-OC, all P_{it} are aggregated to be P, which is only 2388 bits, if $\mathscr{C}$ and Q are 160 bits, respectively. Thus, the communication burden is dramatically reduced by using data aggregation. Besides, the fog aggregates the individual reports to generate a bill for each customer every day. The bill is only 2880 bits. Figure 5.3 illustrates the comparison results of four schemes about the communication burden of SM-to-fog (Fig. 5.3a) and fog-to-OC (Fig. 5.3b). Since the ciphertext of proxy re-encryption [15] is shorter than those of Paillier encryption [27], BGN encryption [18] and Lifted ElGamal encryption [22] (the commitment in [22] is 1024 bits), the communication burden of SM-to-fog in P^{2}SM is lower than those in the other three schemes as shown in Fig. 5.3a. After the individual consumption reports are aggregated, the communication burden becomes constant, which is still lower than those in EPPA [8], Fan14 [25] and Ohara14 [26] (the overhead of fog-to-OC communication is linear to the quantity of smart meters in the home area network). Figure 5.3c shows the comparison result on communication overhead of P^{2}SM and TLS protocol (T-AES) [7], in which AES-256 is utilized to encrypt meter readings and BLS signature [19] is employed to achieve authentication and data integrity. If $N > 2$ in $\mathbb{RA}$, P^{2}SM is more efficient than TLS in fog-to-OC.

Besides, the fog requires enough storage space to transiently store the individual reports in $\mathbb{RA}$. If $N = 1000$ in $\mathbb{RA}$ and $\rho = 15$, these individual reports would possess 30.8 MB storage space every day. Thus, each fog should have 61.6 MB memory to support power consumption collection and dynamic billing for customers.

5.5 Summary

In this chapter, we have defined a new security model to formally describe the misbehavior of fogs and have proposed a privacy-preserving smart metering scheme to balance the end-to-end security and high communication efficiency in smart grid. Not only could P^{2}SM allow fogs to aggregate authentication messages, meter readings and signatures for communication efficiency and privacy preservation for customers, but also prevent a misbehaving fog from corrupting power consumption reports. In addition, the fog can generate verifiable daily electricity bills based on individual consumption reports and dynamic prices. We have proved the security of P^{2}SM and evaluated its performance through the comparison with the existing schemes. P^{2}SM is a secure and efficient communication protocol which is able to replace the TLS protocol to support secure smart metering in smart grid.

References

1. X. Fang, S. Misra, G. Xue, and D. Yang, "Smart grid the new and improved power grid: A survey," *IEEE communications surveys & tutorials*, vol. 14, no. 4, pp. 944–980, 2012.
2. J. Ni, K. Zhang, X. Lin, and X. Shen, "Edat: Efficient data aggregation without ttp for privacy-assured smart metering," in *Proc. of ICC*, 2016, pp. 1–6.
3. J. Ni, K. Zhang, K. Alharbi, X. Lin, N. Zhang, and X. Shen, "Differentially private smart metering with fault tolerance and range-based filtering," *IEEE Transactions on Smart Grid*, vol. 8, no. 5, pp. 2483–2493, 2017.
4. X. Li, X. Liang, R. Lu, X. Shen, X. Lin, and H. Zhu, "Securing smart grid: cyber attacks, countermeasures, and challenges," *IEEE Communications Magazine*, vol. 50, no. 8, 2012.
5. A. Molina-Markham, P. Shenoy, K. Fu, E. Cecchet, and D. Irwin, "Private memoirs of a smart meter," in *Proc. of BuildSys*, 2010, pp. 61–66.
6. F. Cleveland, "Iec tc57 wg15: Iec 62351 security standards for the power system information infrastructure," *White Paper*, 2012.

7. T. Dierks and E. Rescorla, "The transport layer security (tls) protocol version 1.2," Tech. Rep., 2008.
8. R. Lu, X. Liang, X. Li, X. Lin, and X. Shen, "Eppa: An efficient and privacy-preserving aggregation scheme for secure smart grid communications," *IEEE Transactions on Parallel and Distributed Systems*, vol. 23, no. 9, pp. 1621–1631, 2012.
9. H.-Y. Lin, W.-G. Tzeng, S.-T. Shen, and B.-S. P. Lin, "A practical smart metering system supporting privacy preserving billing and load monitoring," in *Proc. of ACNS*, 2012, pp. 544–560.
10. C. Rottondi, G. Verticale, and A. Capone, "Privacy-preserving smart metering with multiple data consumers," *Computer Networks*, vol. 57, no. 7, pp. 1699–1713, 2013.
11. J. Ni, K. Alharbi, X. Lin, and X. Shen, "Security-enhanced data aggregation against malicious gateways in smart grid," in *Proc. of GLOBECOM*, 2015, pp. 1–6.
12. T. Dimitriou and G. Karame, "Privacy-friendly tasking and trading of energy in smart grids," in *Proc. of ACM SAC*, 2013, pp. 652–659.
13. C. Rottondi, M. Savi, G. Verticale, and C. Krauß, "Mitigation of peer-to-peer overlay attacks in the automatic metering infrastructure of smart grids," *Security and Communication Networks*, vol. 8, no. 3, pp. 343–359, 2015.
14. J. Ni, K. Zhang, X. Lin, and X. Shen, "Balancing security and efficiency for smart metering against misbehaving collectors," *IEEE Transactions on Smart Grid*, 2017.
15. G. Ateniese, K. Fu, M. Green, and S. Hohenberger, "Improved proxy re-encryption schemes with applications to secure distributed storage," *ACM Transactions on Information and System Security (TISSEC)*, vol. 9, no. 1, pp. 1–30, 2006.
16. H. Krawczyk and T. Rabin, "Chameleon signatures." in *Proc. of NDSS*, 2000, pp. 143–154.
17. H. Shacham and B. Waters, "Compact proofs of retrievability," in *Proc. of Asiacrypt*, 2008, pp. 90–107.
18. D. Boneh, E.-J. Goh, and K. Nissim, "Evaluating 2-dnf formulas on ciphertexts," in *Proc. of TCC*, 2005, pp. 325–341.
19. D. Boneh, B. Lynn, and H. Shacham, "Short signatures from the weil pairing," in *Proc. of Asiacrypt*, 2001, pp. 514–532.
20. D. Derler and D. Slamanig, "Key-homomorphic signatures and applications to multiparty signatures and non-interactive zero-knowledge," IACR Cryptology ePrint Archive 2016, 792, Tech. Rep., 2016.
21. J. M. Pollard, "Kangaroos, monopoly and discrete logarithms," *Journal of cryptology*, vol. 13, no. 4, pp. 437–447, 2000.
22. T. ElGamal, "A public key cryptosystem and a signature scheme based on discrete logarithms," *IEEE transactions on information theory*, vol. 31, no. 4, pp. 469–472, 1985.
23. B. Libert and J.-J. Quisquater, "Improved signcryption from q-diffie-hellman problems," in *Proc. of SCN*, 2004, pp. 220–234.
24. C.-H. Li and J. Pieprzyk, "Conference key agreement from secret sharing," in *Proc. of ACISP*, 1999, pp. 64–76.
25. C.-I. Fan, S.-Y. Huang, and Y.-L. Lai, "Privacy-enhanced data aggregation scheme against internal attackers in smart grid," *IEEE Transactions on Industrial informatics*, vol. 10, no. 1, pp. 666–675, 2014.
26. K. Ohara, Y. Sakai, F. Yoshida, M. Iwamoto, and K. Ohta, "Privacy-preserving smart metering with verifiability for both billing and energy management," in *Proc. of AsiaPKC*, 2014, pp. 23–32.
27. P. Paillier, "Public-key cryptosystems based on composite degree residuosity classes," in *Proc. of EUROCRYPT*, 1999, pp. 223–238.

Chapter 6
Summary and Future Directions

In this chapter, we summarize the monograph, and discuss several potential research topics for future work.

6.1 Summary

In this monograph, we have introduced the architecture of fog-enabled IoT applications, presented the security and privacy challenges in fog computing, and proposed some promising solutions to protect identity privacy, location privacy and data privacy in fog-enabled IoT applications. Specifically, we have presented the following schemes.

- A privacy-preserving smart parking navigation scheme to preserve drivers' privacy and achieve efficient navigation result retrieval for drivers. The proposed scheme enables a cloud to guide vehicles to vacant parking spaces in the destinations based on real-time parking information, and the fogs to forward the navigation results to the correct querying vehicles without learning any identity information about drivers. In addition, an efficient data retrieval scheme is developed to support navigation result retrieval for querying vehicles. The proposed scheme provides identity privacy-preserving parking navigation with high retrieving probability on navigation results and low computational and communication overhead.
- A fog-assisted privacy-preserving mobile crowdsensing scheme that enables fogs to allocate tasks based on user mobility. The service provider is enabled to allocate the sensing tasks to fogs based on the locations of fogs and the sensing areas of tasks; and thereby the fogs can recruit mobile uses in this coverage areas without having any knowledge about the locations and trajectories of mobile

X. Lin et al., *Privacy-Enhancing Fog Computing and Its Applications*,
SpringerBriefs in Electrical and Computer Engineering,
https://doi.org/10.1007/978-3-030-02113-9_6

users. Therefore, the locations of mobile users are preserved during the task allocation in mobile crowdsensing.

- A privacy-preserving smart metering scheme to prevent pollution attacks and protect meter readings in smart grid. We have defined a new security model to formalize the misbehavior of fogs, in which the misbehaving fogs may launch pollution attacks to corrupt power consumption data. The proposed scheme supports end-to-end security, privacy preservation and integrity protection against the misbehaving fogs under the new security model. It achieves secure smart metering and verifiable dynamic billing against misbehaving fogs with low computational and communication overhead.

6.2 Future Research Directions

This monograph has proposed several privacy preservation schemes for fog computing, including the identity privacy-preserving scheme in smart parking navigation, the location privacy preserving scheme in mobile crowd sensing, and the data privacy-preserving scheme in smart grid. There are other security and privacy issues for fog-enabled IoT applications, which are worth to be studied.

6.2.1 Detection of Rogue Fogs and Devices

As fog computing is faced with a variety of cyber attacks, both fogs and IoT devices contain huge risks at being compromised. The corrupted fogs and IoT devices may pretend to be legitimate users to access services. Even if the fogs and IoT devices are not corrupted, they may become rogue due to their personal incentives. For instance, a rogue fog may be deployed to broadcast rumors and fraud to the drive-by vehicles, or some rogue IoT devices may collude to manipulate the results of mobile crowdsensing. The fake, compromised or rogue fogs and IoT devices would become serious threats to the security and privacy of data. Unfortunately, it is difficult to detect the rogue fogs in fog computing due to the following reasons. Firstly, there are a variety of trust models in different applications, which require distinct trust management schemes. Secondly, the dynamic and distributed environment makes it hard to maintain a blacklist of rogue fogs and IoT devices. Therefore, the study on the methods to detect the rogue and corrupted fogs and IoT devices in fog computing is worthy to focus on.

6.2.2 *Privacy Exposure in Data Combination*

In IoT applications, the devices generate data with various levels of sensitivity on behalf of data producers. Some of these data may be inherently sensitive, e.g., the data generated by a person's heart-rate sensor, but the others might be benign. Nevertheless, even if the data seems to be non-sensitive, the data in combination can trigger serious privacy concerns. Due to the popularity of fogs, collaboration among fogs is often being encouraged, so that the fogs may be able to process data across many IoT devices, which as a result exacerbates the privacy concerns. For instance, a patient buys some pills at a pharmacy and pays using a credit card. The sensitive information of the patient the pharmacy acquires is constrained if the pharmacy does not have the patient's personal information, even if it stores the face and the credit card number of the patient. Nevertheless, if the customers' personal data in local credit card center combines with the purchasing information of patients in the pharmacy, the pharmacy can link the health status with the patient. Consequently, the patient' sensitive information is exposed to the pharmacy. This simple example also demonstrates the importance of identity protection, as benign data may become sensitive, combining with the identity information. The fusion of raw data from different sources can improve the potential values. Therefore, it is necessary to define some levels of privacy protection in data combination and design efficient and effective privacy-preserving schemes to protect users' privacy in fog computing.

6.2.3 *Decentralized and Scalable Secure Infrastructure*

Fog computing is a scalable, dynamic and decentralized paradigm, and it is hard to build a secure infrastructure in such a distributed framework due to the following reasons. First, there is no trusted leader who determines the trustworthiness of fogs. Secondly, conventional security protection approaches are inefficient in fog computing. For instance, to support the authentication of the users' identities and the delegation of access right, each fog needs to maintain a copy of authentication credentials, which results in a heavy storage burden; or the credentials are kept in one powerful fog node, which ends up with heavy communication overhead. It is difficult to find an efficient approach to achieve rapid authentication and delegation in fog computing. Thirdly, although distributed computation can be performed on multiple fogs, it is hard to guarantee the correctness of computation results, since some fogs are not fully trusted. Furthermore, even if these problems can be addressed separately, the compatibility of the solutions may be another critical issue. To build a secure infrastructure in fog computing, one promising technique is to utilize blockchain, which is a distributed ledger that maintains a continuously-growing list of records. The blockchain can build a reliable, trustworthy and powerful architecture for fog computing, but the inherited weaknesses of blockchain should be solved before implementing blockchain in fog computing. In summary, how to build a scalable, efficient and decentralized secure infrastructure is a big challenge but very important for the healthy development of fog computing.

Printed in the United States
By Bookmasters